This book is an essential resource for resilience professionals. Drawing on years of experience and grounded in the diverse coastal Virginia region, it provides an accessible and actionable framework for equitable resilience planning to improve a community's approach to extreme weather and compounding challenges.

Michelle Covi, *Coastal Resilience DOD Liaison, University of Georgia Marine Extension and Georgia Sea Grant*

Anchored by earned trust in the community-led co-production model, and the objective technical expertise of The RAFT team, this field guide draws upon evolutionary best practices in building equitable environmental, social, and economic resilience across vulnerable communities. Building on years of work with coastal localities across a wide range of stakeholders – unique to each – this process ensures authentic unbiased collaboration, essential for a long-term structured planning and adaptive outcomes for use by local and regional communities and governments.

Ann C. Phillips, *Rear Admiral, U.S. Navy (Ret.), former Special Assistant to the Governor for Coastal Adaptation and Protection, Commonwealth of Virginia*

Moving the Needle Towards Resilient Communities

This field guide offers practical guidance for readers to initiate and engage in resilience planning with their communities in a structured, deliberative way that emphasizes equity.

The contents revolve around lessons learned from The Resilience Adaptation Feasibility Tool (RAFT) on-the-ground community initiative to advance resilience across communities in Virginia. Chapters provide background about resilience planning; a starting point for conducting holistic assessment of a community's resilience with a focus on equity; recommendations for establishing an infrastructure for long-term commitment to resilience; and suggestions for identifying and planning activities and projects to build momentum for achieving resilience outcomes.

The format offers an easy read for non-technical audiences such as community planners, community leaders, or nonprofit agency staff to quickly and efficiently get the guidance they need for community-based, equity-focused resilience planning.

Juita-Elena (Wie) Yusuf is Professor of Public Service at Old Dominion University. Her areas of expertise are in public finance, infrastructure planning, coastal community resilience, climate change and sea-level rise, public participation, and stakeholder engagement. She develops resilience products for use by general, non-technical audiences including planners, businesses, and residents.

Tanya Denckla Cobb is Director of the Institute for Engagement & Negotiation at the Weldon Cooper Center for Public Service, University of Virginia. She is a seasoned environmental and public policy mediator, working with people across all sectors to address challenging, complex issues through equitable collaboration and seeking sustainable solutions.

Sierra Gladfelter is Senior Associate at the University of Virginia's Institute for Engagement & Negotiation, where she leads community-engaged climate resilience, facilitation, and collaborative planning projects. Sierra has over a decade of climate adaptation experience and has been the project manager of The RAFT for the last seven years.

Elizabeth Andrews is Distinguished Law & Policy Fellow at the Institute for Engagement & Negotiation, University of Virginia. Her expertise in climate resiliency law and policy derives from many years of teaching and practice with the Virginia Coastal Policy Center at William & Mary Law School and the Virginia Attorney General's office.

Routledge Research in Environmental Policy and Politics

7. **Climate and Clean Energy Policy**
   State Institutions and Economic Implications
   Benjamin H. Deitchman

8. **Populism, Eco-populism, and the Future of Environmentalism**
   James R. Stone Jr.

9. **Partnerships for Sustainability in Contemporary Global Governance**
   Pathways to Effectiveness
   Edited by Liliana B. Andonova, Moira V. Faul and Dario Piselli

10. **The Drought Dilemma**
    States, Innovation, and the Politics of Water Quantity
    Edited by Jonathan D. Farley, Jonathan M. Fisk and John C. Morris

11. **The Meaning of Growth**
    Anti-Environmentalist Rhetoric and the Defence of Modernity
    Richard McNeill Douglas

12. **Moving the Needle Towards Resilient Communities**
    A Field Guide for Equitable, Community-Led Resilience Planning
    Juita-Elena (Wie) Yusuf, Tanya Denckla Cobb, Sierra Gladfelter, and Elizabeth Andrews

Moving the Needle Towards Resilient Communities

A Field Guide for Equitable, Community-Led Resilience Planning

Juita-Elena (Wie) Yusuf,
Tanya Denckla Cobb,
Sierra Gladfelter, and
Elizabeth Andrews

Routledge
Taylor & Francis Group
NEW YORK AND LONDON

First published 2026
by Routledge
605 Third Avenue, New York, NY 10158

and by Routledge
4 Park Square, Milton Park, Abingdon, Oxon, OX14 4RN

Routledge is an imprint of the Taylor & Francis Group, an informa business

ISBN: 978-1-032-97932-8 (hbk)
ISBN: 978-1-032-97933-5 (pbk)
ISBN: 978-1-003-59623-3 (ebk)

DOI: 10.4324/9781003596233

Typeset in Times New Roman
by Newgen Publishing UK

To our colleagues, students, community partners, and quiet champions who have walked beside us through the decade of coastal resilience work of The RAFT: Thank you for your steadfast support, insights, and generous belief in what's possible. This book exists because you never let the tide carry the vision away.

–Wie, Tanya, Sierra, and Elizabeth

For Keegan and Afi, my invisible partners on this coastal resilience journey. For my dad, Yusuf Hadi, whose constant encouragement shaped my passion and strengthened my path.

–Wie

For my soulmate, Cecil, who was the embodiment of resilience and joy.

–Tanya

For Bill, who first showed me how to show up humbly and work hard for a community.

–Sierra

For Ron, my resilient rock, and for all of my students who have committed their time, passion, and inspiring insights to help strengthen RAFT communities.

–Elizabeth

Contents

Figures

Tables

Boxes

Acknowledgments

We thank:

- Tom Allen, Michelle Covi, Trang Le, Karen Firehock and the Green Infrastructure Center, Maya Koehn-Wu, Lauren Buford, Tanya Denckla Cobb, Juita-Elena (Wie) Yusuf, and Virginia Sea Grant for providing the graphics and photographs included in this book.
- The students and staff of the University of Virginia Institute for Engagement & Negotiation, Virginia Coastal Policy Center at William & Mary Law School, ODU-VASG Virginia Climate Adaptation and Resilience Program, ODU Institute for Coastal Adaptation and Resilience, and Virginia Tech Coastal Collaborator.
- Regional planning district commission and locality staff in The RAFT communities.
- Community planners; community-based organizations; leaders of faith organizations, businesses, and local, regional and state agencies; community stakeholders; academic colleagues; and many others who have extensively shown up to support resilience planning and implementation throughout The RAFT process.

We acknowledge the many funders that have enabled this intensive work at the community level:

- Virginia Coastal Zone Management Program, Virginia Department of Environmental Quality and National Oceanic and Atmospheric Administration
- Virginia Environmental Endowment, Virginia Program
- Virginia Sea Grant

- University of Virginia Environmental Institute, formerly Environmental Resilience Institute
- Jessie Ball duPont Fund
- Agua Fund
- Natural Hazards Center, Public Health Disaster Research Award Program
- University of Virginia School of Architecture, Faculty Summer Research Program
- Commonwealth Center for Recurrent Flooding Resiliency
- National Fish and Wildlife Foundation, Restoration and Community Stewardship Grant

1 Setting the stage for equitable resilience

The RAFT approach to an equity-centered path for resilience planning and implementation

Introduction

Resilient coastal communities are able to anticipate threats, reduce their vulnerability to hazards and stressors, and respond to and recover from them more effectively, while remaining economically and socially viable. Being resilient means the community can anticipate, adapt, endure, and thrive in the face of change, uncertainty, and adversity. Over more than a decade, through the development and application of The Resilience Adaptation Feasibility Tool (RAFT), we have worked with coastal communities in Virginia to help them plan for and implement projects to enhance resilience equitably.

This Field Guide is designed to help readers approach resilience planning in a structured way, using a deliberative, community-based process that begins with a holistic assessment of resilience and results in projects that build support for resilience efforts and set the trajectory for coordinated and sustained, long-term resilience. Recognizing that community stakeholders may lack capacity and need additional support, this Field Guide also offers guidance on partnering with organizations such as colleges and universities to support resilience planning.

If you are interested in planning for and enhancing the resilience of your coastal community in ways that consider equity and seek equitable outcomes, this Field Guide is for you. We have written this Field Guide with a wide range of resilience stakeholders in mind.

DOI: 10.4324/9781003596233-1

Box 1.1 Who should read and use this Field Guide?

- Residents and leaders of communities experiencing climate impacts, hazards, and chronic stressors who want to understand how they can be part of efforts to mitigate these impacts.
- Staff and leaders of nongovernmental or community organizations that serve vulnerable or impacted populations, engage in policy and planning efforts that contribute to resilience, or have broad concerns about the resilience of the communities they serve.
- Government leaders and staff involved in activities and functions, such as planning, emergency management, and social services, that are impacted by or contribute to resilience.
- Philanthropists and funders committed to encouraging resilience-building efforts that are driven by community needs, match community priorities, and address equity concerns.
- College or university administrators, faculty, researchers, staff, and students interested in working with and supporting community stakeholders working in the coastal resilience space.
- Other practitioners and community stakeholders invested in the resilience of coastal communities.

This introductory chapter provides the backdrop for resilience planning that emphasizes equity and provides an overview of how we use The RAFT to work with communities of different sizes and geographic challenges, as well as different racial, economic, and other demographics, to help them build their resilience equitably. It highlights our experiences in coastal Virginia and how this experience provides insights and examples of lessons learned for equitable resilience that can be applied in other communities seeking to improve their resilience to social, economic, and environmental challenges.

Defining resilience

Our approach, using The RAFT, defines resilience as the capacity to anticipate threats, reduce a community's vulnerability, and respond to and recover from hazards and chronic stresses. A resilient community is one that can anticipate, adapt, endure, and thrive in the face of change, uncertainty, and adversity. The goal of a resilient community is not just to bounce back from shocks, but also to ensure the well-being of all its members. We view resilience as a multi-dimensional concept that encompasses environmental, ecological, infrastructural, social, economic, and community-oriented needs.

This comprehensive approach to resilience for coastal communities is important for several reasons:

- Addresses disparate impacts and social vulnerability.
  - We explicitly acknowledge that hazards and stressors disproportionately impact low-income populations and racial minorities, threatening their social and economic stability. We recognize that some residents are least able to prepare for, absorb, and recover from shocks, such as those in older and poorer neighborhoods or those facing financial instability.
  - This approach highlights the need to ensure all voices are heard and to prevent or address disparate impacts and varying abilities to cope with hazards.
- Integrates economic and social viability.
  - Under-resourced communities may view economic and social resilience as more urgent priorities than coastal hazards. Our approach connects the dots between coastal resilience and broader concerns, including physical and mental health, and access to care, jobs, and financial resources.
- Fosters a comprehensive and inclusive planning approach.
  - By defining resilience as multi-dimensional, we encourage a whole-of-community approach that involves various sectors like government, business, non-profit, academic, faith-based, and civil society. This multi-sectoral approach helps overcome policy and sector silos and allows for integrated resilience planning.
  - We recognize that many community planning efforts are vital tools that can be coordinated and repurposed to address a broad range of challenges, including social, economic, and environmental issues for coastal communities.

In essence, The RAFT's definition of resilience is important because it embeds equity, social justice, and holistic community well-being at its core, thereby promoting a more sustainable and inclusive path for adapting to coastal hazards.

Resilience planning

Resilience planning is a comprehensive and multifaceted process that communities can undertake to proactively and strategically build resilience. Resilience planning is an iterative process that moves beyond merely reacting to hazards and stressors to proactive preparation and long-term sustainability. It is fundamentally important for communities to anticipate and respond to disruptions, thereby reducing vulnerability and ensuring that essential systems continue to operate or quickly recover from hazardous events and chronic stresses. Effective resilience planning relies on both quantitative and qualitative information to inform planning and policy decisions. Moreover, it promotes holistic community well-being by integrating environmental, ecological, infrastructural, social, and economic aspects into the community's resilience efforts. It also fosters integration and coordination among various planning efforts and stakeholders, avoiding counterproductive redundancies that consume limited resources and preventing fragmented efforts that could increase vulnerability and detract from resilience.

Equity in resilience and resilience planning

Equity is a moral imperative and a strategic necessity in resilience planning, crucial for ensuring that efforts benefit all community members, especially the most vulnerable. Equitable resilience planning ensures that all voices are heard and addresses disparate impacts and coping abilities. Hazards and stressors disproportionately impact low-income populations, racial minorities, and those facing financial instability. Without an explicit focus on equity, planning processes can exacerbate existing injustices by favoring privileged groups and further marginalizing vulnerable communities. Addressing equity concerns also ensures that planning extends beyond physical infrastructure to encompass crucial social and economic dimensions. It focuses on meeting the basic needs of all residents as a fundamental step in building overall resilience.

To accomplish this, listening to those who are impacted is most important. While it may not be complicated, this effort does require

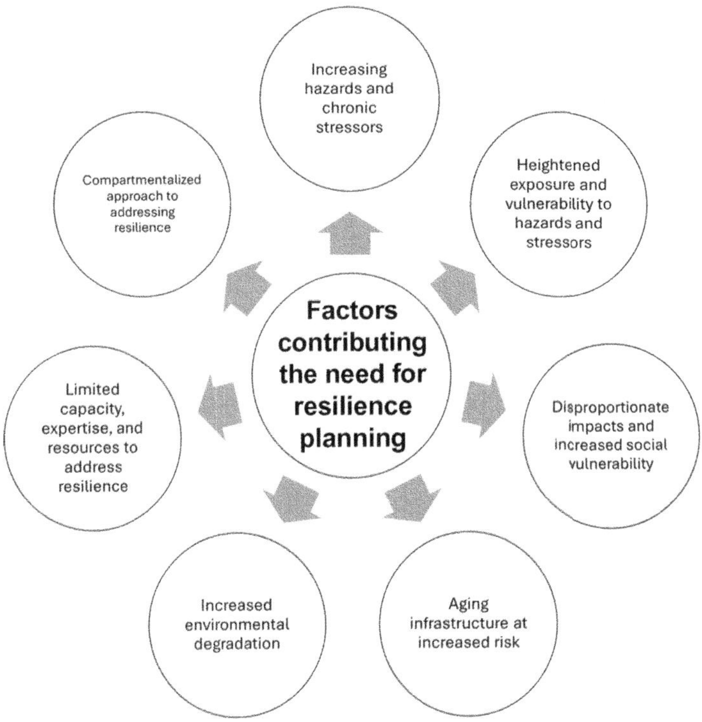

Figure 1.1 Factors contributing to the need for resilience planning.
Source: Created by authors.

time, attention, and explicit inquiry before decisions are made, at the earliest possible time, about a host of key issues, such as:

- Their key concerns and hopes.
- The information they need to participate meaningfully and in what language(s).
- The source(s) of information they trust.
- Their preferred method of communication.
- Where and when meetings should be held to enable their effective participation.
- Whether they need transportation, child care, and/or food.

Perhaps most important for equitable engagement is the need to be clear and precise about what specifically the community's input can and cannot influence. There is no surer way to create distrust and anger in a community than to ask for community input on an issue and then ignore it. To build trust and relationships across the community, work with integrity. Say what you'll do, and do what you say you'll do.

The RAFT and community-driven, equity-focused resilience planning

The Resilience Adaptation Feasibility Tool (RAFT) is an innovative, collective impact, collaborative approach to climate resilience that is designed to help communities, particularly coastal localities, improve their resilience to flooding and other coastal hazards while remaining economically and socially viable. The RAFT recognizes that resilience is a multi-dimensional construct, encompassing environmental, ecological, infrastructural, social, and economic aspects. It is both a tool for resilience planning and project implementation and a process for ensuring equitable resilience. The RAFT design and application inherently embody the principles and conditions of collective impact, particularly through a process emphasis, including broad collaboration, a shared vision for equitable resilience, and a structured, adaptive approach.

Box 1.2 Why collective impact?

Collective impact is a structured approach that brings people together to achieve social change. It involves a network of community members, organizations, and institutions working collaboratively to advance equity by learning together, aligning actions, and integrating efforts to achieve population and systems-level change.

Five core conditions characterize collective impact initiatives:

- **A common agenda**: Participants collectively define the problem and create a shared vision to solve it.
- **Shared measurement**: Progress is tracked in the same way, enabling continuous learning and accountability.

- **Mutually reinforcing activities**: The diverse activities of participants are integrated to maximize the overall result.
- **Continuous communication**: Frequent and transparent intra-network communication fosters trust and strengthens relationships among collaborators.
- **A strong backbone**: A dedicated team is responsible for aligning and coordinating the work of the entire group.

Furthermore, collective impact efforts explicitly center equity. This is achieved through five strategies that should be woven into the implementation of the core conditions:

- Grounding the work in data and local context and targeting solutions.
- Focusing on systems change, in addition to programs and services.
- Shifting power to transform the collaborative from doing things *for* a community into sharing authority *with* the community.
- Listening to and acting with the community.
- Building equity leadership and accountability.

The RAFT is a multi-university, interdisciplinary academic partnership initiated in 2015. Our core team initially included the Institute for Engagement & Negotiation (IEN) at the University of Virginia, the Virginia Coastal Policy Center (VCPC) at William & Mary Law School, and the Old Dominion University (ODU) and Virginia Sea Grant Virginia Climate Adaptation and Resilience Program. Our team has grown over the years and now includes ODU's Institute for Coastal Adaptation and Resilience and the Virginia Tech Coastal Collaborator.

The creation of The RAFT was a direct response to the increasing need for coastal communities across Virginia to enhance their resilience to coastal hazards. The idea for The RAFT originated from a 2015 meeting convened by IEN and VCPC staff that brought together government and community stakeholders in coastal Virginia localities to explore how to increase resilience. At this meeting, stakeholders identified several barriers to resilience, including a lack of clear understanding of vulnerabilities and opportunities, confusion about solutions, and limited local government capacity to manage resilience initiatives. The meeting also highlighted the need for a report card that could be used to comprehensively assess a locality's resilience in an easily understandable format for community leaders. Originally conceived by some advocates as a way to incentivize competition among local governments to improve their resilience, the idea of publicized grades on the report card was quickly abandoned in favor of a scorecard intended to educate and guide local governments towards concrete actions. This led to the joint development of a coastal resilience scorecard, marking the beginning of The RAFT partnership.

To develop The RAFT's resilience scorecard, we conducted an in-depth analysis of existing national and international resilience and sustainability scorecards and report cards. An advisory committee of over 20 subject matter experts from government, nonprofit, and academic sectors was organized to provide guidance and feedback that informed the development of the scorecard. We held two focus groups with these subject matter experts. One focus group, involving staff from state and local government agencies, focused on determining the relevant components to include in the scorecard and evaluating its objectivity. The feedback from this focus group emphasized the importance of an academic team conducting the analysis using the scorecard to ensure objectivity and neutrality. Another focus group was held with social equity experts to address how equity could be incorporated into the scorecard. The social equity experts advised that equity should be integrated throughout the scorecard rather than treated as a separate section. They also helped develop key metrics for identifying, mapping, informing, and engaging socially vulnerable populations, as well as for working with their trusted communication messengers and networks.

The resulting RAFT Scorecard incorporated physical risk, as well as political, economic, and social factors, assessing resilience across five dimensions: policy, leadership, and collaboration; risk assessment and emergency management; infrastructure resilience; planning for resilience; and community engagement, health, and well-being. The scorecard was pilot-tested in three diverse Virginia localities – the City of Portsmouth, Gloucester County, and the Town of Cape Charles – representing urban, suburban, and rural settings and differences in demographics, municipality type, density, community size, and physical vulnerability. Feedback from the pilot testing led to

Figure 1.2 Community stakeholders participate in identifying and prioritizing resilience action projects.

Source: Photograph provided by Tanya Denckla Cobb. Used with permission.

further integration of equity across The RAFT Scorecard, including greater emphasis on vulnerable populations and community-wide communications.

While the initial impetus for The RAFT was the need for a scorecard to assess resilience strengths and gaps in policies and plans, the research into existing resilience and sustainability scorecards and report cards, and interviews with developers of five of the most promising models, revealed that the mere act of scoring was not sufficient to lead to change or actions on the ground. Instead, support for communities was needed to help them translate the findings of the scorecard assessment into specific actions, to address their gaps in policies and planning for resilience. The result was an expansion of The RAFT process to include two additional steps: a workshop for the community to develop specific resilience priority actions to address their policies and planning gaps; and a year-long phase of support for the locality's implementation of these priority actions.

The RAFT process, as initially conceived, involved three main phases over 18 months:

1) **Scorecard assessment (Phase 1)** – We conduct an independent assessment of a locality's resilience that is verified with locality staff throughout the assessment process to ensure accuracy and to avoid surprising staff with assessment results when they are presented to elected officials. This brings academic objectivity to the process, reducing perceived political influence and allowing for more authentic consideration of equity in community discussions. Assessment results are presented to local government staff and elected officials, providing a clear diagnostic picture of existing strengths and opportunities for improvement. This assessment helps localities understand the current resilience landscape and potential solutions.
2) **Development of a Resilience Action Checklist (Phase 2)** – Following the assessment, we convene a regional community workshop, co-hosted with the regional planning organization, that brings together a broad range of stakeholders, including local, regional, and state government representatives, non-profits, faith-based organizations, educators, community leaders, and residents. This workshop is intended to facilitate community conversations to identify and prioritize resilience projects. This workshop employs a co-production approach using presentation of localized scientific information and participatory processes to collaboratively develop a Resilience Action Checklist. The aim is to integrate scientific knowledge, public perceptions, community concerns, and institutional capabilities into a prioritized list of resilience projects. Workshop participants prioritize three to five achievable actions that can be completed or significantly progressed within a one-year period. The workshop prompts participants to brainstorm actions that address equity issues unique to their community, such as those related to social vulnerability, health, and wellness, making equity central to resilience considerations.
3) **Implementation of resilience action priorities (Phase 3)** – The RAFT recognizes that rural localities and small cities with socio-economic challenges often have limited capacity (staff, technical expertise, and financial resources). We aim to bridge these gaps by providing structured processes, technical assistance, and administrative support to reduce the administrative burden on locality staff. The RAFT provides one year of ongoing support for implementing the prioritized projects on the Resilience Action Checklist. This includes facilitating regular meetings (virtual or in-person) with

an Implementation Team to monitor progress and provide support. A key part of this phase is connecting the dots for services and programs that work with vulnerable populations. The RAFT helps connect service agencies (e.g., social services, health departments) into the coastal resilience network to meet the needs of vulnerable populations and address broader social and economic resilience needs, such as food access or transportation, which are often overlooked in traditional coastal resilience planning. Examples include connecting community leaders with technical experts for food security plans or facilitating partnerships for shoreline erosion projects. This phase also seeks to balance early wins with long-term strategic impacts to maintain momentum and engagement. The implementation phase concludes with a regional wrap-up workshop for communities to share progress and learn from each other.

The RAFT deployment and continuous growth

After pilot testing and refinement, we deployed The RAFT in 2018 to support resilience planning and implementation in seven localities on the Eastern Shore of Virginia. We met regularly with regional planning staff, locality staff, elected officials, community leaders, and residents to implement The RAFT in these communities. In 2019, we began resilience assessments for eight localities in the Northern Neck region of Virginia. However, due to the COVID-19 pandemic, we shifted from a regional workshop to holding individual community workshops virtually via Zoom and supporting the Implementation Teams via monthly Zoom meetings. As we saw the effects of the pandemic on communities across coastal Virginia and beyond, we recognized the importance of supplementing The RAFT Scorecard with a qualitative assessment.

We conducted our first qualitative assessment in the Northern Neck through interviews and listening sessions with community service providers to better understand the COVID-19 pandemic impacts and repercussions for resilience. We have since expanded The RAFT's resilience assessment to include both The RAFT Scorecard (quantitative assessment) and a qualitative assessment that captures local leaders' views on challenges, strengths, and opportunities for building resilience. We have also modified The RAFT resilience planning and implementation process by more intentionally broadening the invitation list for community workshops to ensure representation of socially vulnerable groups and enhancing Implementation Teams with community volunteers, a wide range of local government departments, staff of relevant state agencies and community and non-profit organizations, and faculty providing support for specific implementation projects.

In 2021, we partnered with the Middle Peninsula Planning District Commission to implement The RAFT in six localities across Virginia's Middle Peninsula. Additionally, as we continued working with coastal communities, we recognized that coastal hazards, such as flooding, were not the only threats to which these communities were vulnerable. First, we saw how riverine and inland flooding posed challenges to many coastal-adjacent communities. Second, we realized that many communities have populations that are vulnerable to extreme temperatures, and particularly increasing heat. We subsequently revised The RAFT Scorecard to incorporate specific concerns around extreme heat and cold and riverine and inland flooding, expanding its applicability beyond coastal hazards. We used this modified RAFT Scorecard in 2023 to support two mountain communities in southwest Virginia. With these two localities, The RAFT process helped address broader resilience goals such as heat relief, mobility, and economic diversification.

More recently, we further refined and expanded The RAFT Scorecard to account for tribal governance variances and increase its utility for tribal nations. This revised scorecard and a modified resilience planning process have provided a successful framework for supporting four Virginia Indian Tribes. The modified process reflected the need to respect tribal sovereignty and the proprietary nature of the information used in their internal resilience assessment and development of Resilience Action Checklists.

As just described, The RAFT works with communities of different sizes, geographic challenges, and racial, economic, and other demographics, with diverse needs and cultures. Applications of The RAFT have extended to various coastal and coastal-adjacent localities, tribal nations, as well as mountain communities across Virginia, encompassing 35 communities by 2026. Through this progression, and learning what may be important to different communities, we have learned that climate resilience – or even just resilience – is more broadly applicable and understood than coastal resilience. Similarly, as we are working with tribes and mountain communities, we have shifted from speaking about coastal hazards to climate and other hazards. Importantly, we have adapted The RAFT approach to suit the unique contexts of these communities. We have found it is essential to work closely with the regional planning agency, as a way of assuring that this form of university support is desired, welcomed, and will be useful. Part of engendering this welcome is the promise that we will tailor The RAFT application to account for different regional needs and priorities. Another factor is that this work requires being sensitive to the political settings and landscapes of the different regions and learning what role the regional planning organization wishes to play in introducing and participating in The RAFT process, as well as what role the local governments wish to have during the year of implementation. For example, in rural localities, where staff capacity, technical expertise, and financial resources are often limited, The RAFT may be needed to provide administrative support, in addition to connecting the dots between resilience and broader socio-economic concerns like physical and mental health and access to care, jobs, and financial resources. This flexibility in our approach recognizes that under-resourced communities may prioritize immediate economic and social concerns over climate hazards, thereby framing hazards within the broader context of community well-being. The RAFT aims to move beyond superficial equity considerations to recognize deeper complexities of social interactions and injustices, acknowledging interdependencies between acute shocks (e.g., climate hazards) and chronic stressors (e.g., poverty, aging infrastructure, and racial injustice). For example, while many resilience practitioners may not see housing as an important piece of the climate resilience puzzle, the aftermath of devastating hurricanes or other stressors can quickly demonstrate how a lack of transitional, attainable housing can be one of the greatest barriers to recovery.

How does The RAFT support localities?

The RAFT is both a tool and a process for resilience planning and project implementation that brings a multi-level approach to building resilience – centering equity, recognizing that resilience begins with individuals, households, and businesses, and working directly with localities within a regional context.

The RAFT's academic partnership brings objectivity to the resilience assessment, prioritization of actions, and project implementation, which helps to reduce perceived political and governmental influence and allows for more authentic consideration of equity in community discussions and actions. The RAFT's work is supported by a mix of federal, state, and private foundation grants and donated services, provided at no cost to participating localities and tribal nations. In this way, The RAFT is neither beholden to nor a program of the local government whose programs and policies are being assessed.

We collaborate with regional planning organizations, local governments, community organizations, and other stakeholders. This regional approach ensures a community-led focus for developing and implementing resilience actions, while also enabling localities to discuss and cooperate on shared goals, learn from one another, and avoid duplication of effort.

The RAFT also recognizes that hazards and stressors disproportionately impact socially vulnerable populations, including low-income individuals, racial minorities, older people, children, and people with disabilities, who often lack sufficient resources to cope with disruptions. Equity is thus integrated throughout The RAFT process, helping to ensure that vulnerable populations' needs are taken into account.

In essence, The RAFT supports localities by providing a structured, comprehensive, and equitable framework for resilience planning that is community-led, data-informed, and capacity-building. Moreover, The RAFT process addresses not only environmental risks but also crucial social and economic vulnerabilities. It continually evolves to meet community needs, such as expanding its scope to include riverine flooding and extreme temperatures, and more fully integrating social equity efforts.

The RAFT's 18-month engagement with localities is a deliberate choice to ensure that resilience planning is not merely a technical exercise but a deeply embedded, community-driven process that leads to concrete actions to address complex, interconnected vulnerabilities in an equitable, sustainable, and actionable manner. This timeline is rooted in several critical considerations for effective and equitable resilience building:

- **Undertaking comprehensive data collection and analysis.** The quantitative and qualitative assessments require sufficient time to gather and analyze a wide array of documents, ordinances,

Figure 1.3 Participants in The RAFT community workshop reviewing resilience action projects.

Source: Photograph by Trang Le. Used with permission.

and policies, as well as conduct in-depth interviews and listening sessions with diverse community members and experts.

- **Building trust and engagement for planning.** The qualitative assessment and workshop phases require time to build relationships and trust among diverse stakeholders, especially with historically underserved or excluded groups. This is important for ensuring that all voices are heard and that the resulting Resilience Action Checklist is truly co-produced and reflective of community priorities, not just top-down mandates.
- **Achieving tangible progress and momentum.** A full year of dedicated implementation support allows communities to make meaningful progress on identified actions and build the networks to sustain initiatives, leading to early gains that can build momentum for long-term resilience efforts. Short-term efforts might not fully address the complex issues associated with resilience.
- **Engaging in capacity building and resource mobilization.** Many rural localities that The RAFT primarily serves have limited staff capacity, technical expertise, and financial resources. The year-long

implementation phase allows different community groups who may never work together to experience how they might collaborate in unexpected ways for a common cause, helping break down silos that exist even in the smallest communities. This length of time also enables us to provide ongoing administrative and technical support, helping localities to mobilize existing capacity and fill capacity gaps. This sustained assistance helps overcome common barriers to resilience action.

- **Fostering adaptive management and learning.** Resilience planning is inherently an iterative process that requires continuous learning and adaptation to changing conditions and new information. A 12-month implementation period allows for regular monitoring, adjustments to strategies, and the integration of lessons learned, which are crucial for building long-term adaptive capacity.
- **Broadening community involvement.** Sustained engagement over a year-long period is necessary to broaden participation beyond traditional stakeholders and to ensure that less visible or less engaged vulnerable populations are reached and supported. The locality Implementation Teams expand throughout the year of implementation, as additional participants are identified and invited to join. This helps address social equity by promoting procedural justice, recognizing diverse identities and needs, and enlarging the voice of vulnerable communities in local resilience discussions, hopefully beyond the year of implementation.

Why The RAFT is the basis for this Field Guide

This Field Guide is based on our RAFT team's experiences over the last decade, which have provided insights, lessons learned, and case examples for equitable resilience planning and project implementation. Several features of The RAFT offer a strong basis for resilience planning for communities beyond those in Virginia.

First, The RAFT is an iterative process for communities to increase their resilience to hazards and stressors given the growing emphasis on being resilient, as well as the need to keep pace with change. These include changes in the local natural, physical, cultural, social, and economic environments due to rapid development, a changing climate and economy, environmental degradation, shifting community and population demographics, and other factors.

Second, The RAFT recognizes that there is no one-size-fits-all approach to resilience planning and implementation. Different communities will have unique resilience needs that emerge from the comprehensive assessment. The projects identified on the Resilience Action Checklist are selected and prioritized by the community to meet their specific needs while recognizing their constraints.

Third, we rely on a co-production approach to collaboratively assess resilience, develop the Resilience Action Checklist, and implement resilience projects. This approach integrates scientific knowledge, public perceptions, community concerns, and institutional capabilities. We learned the importance of intentionally broadening invitation lists for workshops and Implementation Teams to include members of socially vulnerable groups and community volunteers, recognizing that local governments may not have the capacity or trust to build intentional relationships with vulnerable populations. This is consistent with longstanding best practices to seek input from participants in designing how they participate in public engagement (see Box 1.3 for more information).

The RAFT recognizes that community members play different roles that contribute to the shared success of resilience planning and implementation. Regular workshops and meetings, such as the community workshop to develop the Resilience Action Checklist and meetings with the Implementation Team, provide consistent interactions that help build trust and shared understanding. We have learned the importance of having local partners or consultants with connections to or relationships with the community, especially in tight-knit communities. Because resilience planning is not a one-and-done approach, continuity and support within the community are needed, along with early and sustained engagement with community members. However, we also acknowledge the challenge that long-term engagement, while needed, is difficult without sustained or targeted funding.

Box 1.3 Co-production and meaningful public participation

Co-production is a collaborative approach where researchers, practitioners, community members, and/or other stakeholders jointly create knowledge, tools, and solutions, such as those addressing resilience challenges. Co-production positions community members as equal partners who can provide lived experience, local knowledge, and valuable insights. This approach helps ensure that resilience strategies are grounded in real community needs, concerns, and capacities, increasing their relevance, legitimacy, and long-term sustainability.

The International Association for Public Participation (IAP2) emphasizes that meaningful engagement, such as through co-production, requires more than simply inviting people to participate. The IAP2 Core Value #5 highlights the importance of seeking input from participants themselves. This means co-designing the process (e.g., meeting formats, communication strategies, and decision-making structures) in ways that reflect community preferences and reduce barriers to participation.[1] Doing so particularly benefits socially vulnerable groups who may otherwise be excluded due to institutional constraints or historical mistrust.

Sherry Arnstein's classic article "A ladder of citizen participation" illustrates the spectrum of public involvement, from nonparticipation to tokenism (informing, consultation, placation) to genuine citizen power (partnership, delegated power, citizen control).[2] The ladder emphasizes that not all engagement is equal, and moving toward higher rungs requires sharing power and enabling communities to influence decisions that affect them. Co-production aligns with the upper levels of the ladder, where residents are not simply consulted but become partners in shaping resilience priorities and actions.

Fourth, resilience planning and implementation needs to be adaptable to the community context and evolve as things change, mistakes are made, and lessons are learned. With The RAFT, we have emphasized the need to continuously evolve as we work with different communities. The adaptability and broader applicability of the underpinning framework point to how The RAFT can be used for resilience planning and implementation for various contexts.

Organization of this Field Guide

This Field Guide is organized around six success factors we have identified from our work with communities using The RAFT.

1) Equity, as both a process and an outcome, is at the heart of the community's resilience.
2) Resilience planning begins with a holistic assessment of resilience that is customized to the community's needs.
3) Even small projects can have outsized impacts as catalysts for action and sustaining momentum for long-term resilience.
4) Resilience spans different facets of community planning, and equitable resilience can be embedded within existing planning processes.
5) Resilience connects across the whole community, and sustained progress requires coordinating across partners and stakeholders involved in resilience efforts.
6) Federal and state agencies, nonprofits, extension services, colleges and universities, and other organizations can support resilience efforts, but they often need guidance in how they can help.

These success factors are explained in the following chapters. The chapters are organized to introduce and explain the critical success factors, lessons learned, and concrete examples from our experience supporting community-based, equity-driven resilience planning in Virginia, along with reflections on how communities can be more intentional and deliberate in planning for resilience and implementing projects that increase it. The chapters in this Field Guide offer a starting point for readers engaging in efforts to plan for and enhance their community's resilience. Each chapter is a standalone chapter. You do not need to read the chapters in the order they are presented in this Field Guide.

Notes

1 International Association for Public Participation USA. (n.d.) *IAP2 Core Values, Ethics, + Spectrum.* IAP2 USA. https://iap2usa.org/cvs

2 Arnstein, S.R. (1969). A ladder of citizen participation. *Journal of the American Institute of Planners*, 35(4), 216–224.

2 Centering equity in resilience

Equitable resilience as a process and an outcome

Introduction

Resilience is not uniformly distributed throughout a community, and hazards and stressors can disproportionately impact socially vulnerable populations that lack sufficient resources or capacity to cope. Traditional resilience thinking, which expects communities to bounce back, can be problematic when applied to areas affected by generations of systemic inequities.[1] This approach is particularly flawed in the context of unfair land-use policies, such as the deliberate creation and maintenance of urban ghettos and areas of concentrated poverty through discriminatory zoning, restrictive covenants, and government disinvestment.[2,3] Without a focus on equity, resilience efforts risk exacerbating existing injustices and disparities and failing to meet the needs of all residents. The increasingly complex challenges of climate change, coastal hazards, and chronic social and economic stressors necessitate a structured process and specific tools to support planning and implementation efforts for equitable resilience.

The Resilience Adaptation Feasibility Tool (RAFT) directly addresses this need by providing a comprehensive, community-driven framework that integrates equity into all phases of resilience planning and project implementation. This chapter explains how and why building resilience requires an equitable process to achieve equitable outcomes. We define equitable resilience and characterize inequities that can undermine it across a community. We introduce the principles that underpin The RAFT's approach to equitable coastal resilience and offer examples of how The RAFT has addressed equity and resilience challenges in communities.

DOI: 10.4324/9781003596233-2

Equity challenges to resilience

Inequities often exist across communities, stemming from historical injustices, socio-economic disparities, and uneven distribution of resources, all of which pose substantial challenges to building resilience. Planning and development approaches can favor some privileged groups at the expense of marginalized communities, entrenching speculative, exclusionary, or unsustainable practices.[4] For example, the use of cost-benefit analyses may perpetuate reliance on expensive grey infrastructure (e.g., sea walls or levees) to protect the economic core (e.g., downtown business districts and high-value waterfront properties) rather than on nature-based or retreat strategies that could benefit a broader, more vulnerable population. Such an entrenched, privileged practice favors the concentration of capital over equitable risk reduction.

Environmental justice issues continue to persist, with low-income neighborhoods and historically underserved populations lacking resources to deal with issues such as flooding. Residents in older and poorer neighborhoods, those facing financial instability, and racial minorities are least able to prepare for, respond to, and recover from hazards. Low-income populations and racial minorities often disproportionately bear the brunt of hazards like recurring flooding, which undermines property values and threatens economic stability. Social vulnerability is increased by conditions such as high unemployment, poverty, medical fragility, and social isolation. Specific groups like women, children, older people, people with disabilities, and minorities are over-represented in vulnerable populations.

Uneven access to benefits and resources can pose challenges for resilience. There are documented inequities in flood insurance coverage, property buyouts, siting of stormwater infrastructure, and home reconstruction efforts during recovery.[5] In some cities, historically Black and Latino neighborhoods have been shown to have substandard stormwater infrastructure, while affluent, whiter neighborhoods are typically served by higher-capacity, underground storm sewer systems.[6] Additionally, history shows that large infrastructure projects (such as highways and industrial facilities) were deliberately sited through or adjacent to communities of color. Communities of color and low-income areas are less likely to have their flood risk accurately mapped,[7] meaning that property owners are often not required to purchase flood insurance, leading to lower insurance take-up rates and no financial safety net when they inevitably flood.[8] Additionally, studies have shown that households in census tracts with a higher proportion of people of color receive less on their claims and have a higher rate of unpaid claims than similar households in predominantly white tracts.[9, 10] This disparity is tied to the systemic undervaluation of property and possessions in low-income and minority neighborhoods, which limits the declaration of a major disaster, eligibility for federal individual assistance, and maximum allowable claim payout.

More broadly, affordable housing and emergency shelters are often limited, impacting vulnerable populations, including a largely invisible homeless population. In many communities, there are pockets of neighborhoods with limited access to essential services such as medical and mental health services, healthy food, and pharmacies. A qualitative assessment conducted by our team during the COVID-19 pandemic identified a lack of broadband access and low levels of computer literacy as critical barriers to employment and education. Transportation and connectivity issues are significant in rural areas, hindering access to services and employment. Language barriers can prevent people who speak English as a second language from accessing resources.

Additional concerns emerge from power imbalances and limited participation in resilience-related efforts. There is frequently a lack of diversity in representation on government and non-profit boards and councils, limiting the inclusion of diverse perspectives. Low levels of public awareness about resilience options and few efforts to engage stakeholders on these issues create inequities in knowledge and support. Traditional community engagement processes often fail to meaningfully include marginalized groups, instead attracting unrepresentative voices and leading to imbalanced power dynamics. Vulnerable community members often do not trust that they will be heard in public meetings and may find it hard to attend meetings held at inconvenient locations or times. This results in marginalization of less advantaged groups and their needs in formal decision-making and planning processes.

When community participation in decision-making about resilience is ignored, the sustainability and legitimacy of plans and their implementation can be called into question. Without early and sustained

Figure 2.1 As some residents elevate their homes above floodwaters, others are left at greater risk, highlighting the need for equitable resilience solutions.

Source: Photograph provided by Virginia Sea Grant. Used with permission.

stakeholder involvement, community members may feel frustrated, and distrust in the planning process can reduce public support for plans and projects.

These inequities challenge local government efforts to build resilience by:

- Increasing the vulnerability of certain populations and exacerbating existing disparities, leading to an uneven distribution of harm and recovery capacity.
- Disempowering local communities by imposing solutions defined externally without sufficient community agency and self-determination.
- Contributing to a lack of trust and social cohesion, which are critical elements for community resilience and collective action.
- Undermining the legitimacy and effectiveness of resilience plans and projects that do not reflect the diverse needs and priorities of all community members.

- Leading to maladaptive solutions where resilience projects implemented for some groups may inadvertently increase vulnerability or entrench injustices for others.

What does equitable resilience look like as an outcome? Equitable resilience, as an outcome, moves beyond simply bouncing back from a shock to intentionally bouncing forward toward a more just, sustainable, and inclusive state for all community members. Several characteristics can be seen as features of equitable resilience, including:

- Reduced and equally distributed vulnerability where the most vulnerable populations and the most privileged have the same protective resources and adaptive capacity.
- Just and inclusive recovery that provides equitable access to aid to current residents (to avoid displacement) and reinvests in community assets.
- Authentic community self-determination with increased social cohesion and trust based on decision-making processes that ensure diverse representation and meaningful participation.

In essence, equitable resilience is an outcome where the entire community has the capacity not just to survive disruptions, but to emerge from them healthier, safer, and more just than before.

Building resilience with an equitable process to achieve equitable outcomes

Equitable resilience is both a moral duty and a strategic necessity for a community. However, equity is also a process, not just an outcome. It results from processes and practices that consider and account for issues of social vulnerability and differential access to power, knowledge, and resources. Equitable resilience entails ensuring that all voices are heard, addressing disparate impacts and coping abilities, and ultimately increasing resilience equitably. For example, an equitable process explicitly recognizes how conditions such as high unemployment, lower incomes, poverty, medical fragility, and social isolation increase social vulnerability. When an equitable process is not followed, these most vulnerable groups may experience greater harm and/or have lesser capacity to cope. Without an equitable process, resilience efforts risk becoming superficial or even counterproductive, deepening existing disparities rather than alleviating them.

An equitable process is essential for achieving equitable resilience outcomes for several reasons and in various ways:

1) **Fostering community-based and inclusive approaches** – Equitable resilience planning is community-based and inclusive of diverse community members, especially those historically underserved and excluded. This involves a process that integrates scientific knowledge, public perceptions, community concerns, and institutional capabilities to develop solutions collaboratively among policymakers, government officials, businesses, faith-based organizations, non-profits, community leaders, and residents. Engaging diverse participants brings unique knowledge, risk perceptions, expertise, and resources to the table, helping to overcome "acts of commission" (decisions disproportionately affecting disadvantaged groups) and "acts of omission" (plans that leave out disadvantaged groups' needs or protect the privileged at the expense of others).[11]
2) **Addressing interdependencies and chronic stressors** – Communities often have broader social and economic concerns (such as food security, health care access, jobs, and financial resources) that are deeply intertwined with coastal resilience. Equitable resilience requires a systems approach that acknowledges the interdependence between acute shocks (e.g., coastal hazards) and chronic stressors (e.g., poverty, aging infrastructure, racial injustice). It involves moving beyond a sole focus on physical infrastructure to encompass broader socio-economic concerns like physical and mental health, access to medical care, jobs, and financial resources. This interconnectedness requires multi-sectoral and whole-of-community approaches that span government, business, non-profit, academic, faith-based, and civil society sectors. Effective planning must "connect the dots" between coastal hazards and other fundamental needs.
3) **Overcoming fragmented planning and policy silos** – Traditional planning efforts are often fragmented, with various departments and organizations operating in silos, leading to weak coordination, conflicting policies, and increased vulnerability. This can result in maladaptive outcomes that favor privileged groups and further marginalize disadvantaged communities. For example, land-use restrictions and infrastructure investments can conflict, with planning agencies aiming to limit development in high-risk areas while infrastructure departments are extending investments

into these zones in a way that subsidizes development and increases flood risk. This dynamic displaces vulnerable residents and entrenches inequitable, unsustainable practices. A structured, equity-focused approach emphasizes integration to overcome fragmentation and silos.

4) **Continuous learning and adapting** – Building resilience is an ongoing process, not a destination. It requires continuous engagement, learning from experiences, and adjusting plans based on new information and changing conditions. This adaptive management approach is critical because communities face complex, uncertain, and unpredictable changes.

The RAFT equitable resilience planning and implementation framework

The RAFT was designed to support localities in creating pathways to incorporate equity into coastal resilience. A focus on equity as an outcome and a process is integrated throughout The RAFT's three-phase process.

1) **Resilience assessment** – The RAFT Scorecard assesses resilience across five dimensions, with social equity explicitly included in "community engagement, health, and well-being" and integrated throughout all other dimensions. This includes evaluating whether risk assessments are conducted for vulnerable populations and if meaningful emergency information is provided to them. For example, metrics include whether a locality or tribe provides coastal resilience information through multiple channels (social media, website, public meetings) and whether emergency alerts and preparedness resources are accessible to all residents and tribal members, including vulnerable populations. Qualitative assessments, through interviews, surveys, and listening sessions with government staff, community service providers, and community leaders, further illuminate challenges and opportunities from the on-the-ground perspectives of those who work with historically excluded and underserved populations. This assessment provides data to identify equity gaps and concerns, informing resilience planning and implementation efforts. (See Table 2.1 for a list of example questions asked about equity as part of The RAFT qualitative assessment.)

2) **Co-development with community stakeholders of a Resilience Action Checklist** – The RAFT process includes convening regional community workshops to bring together diverse stakeholders, including staff of government agencies, non-profits, faith-based and community organizations, and residents. A co-production approach is used to collaboratively develop priority actions, ensuring equity is front and center. For example, workshop participants are asked to brainstorm ideas that address unique equity issues, such as those related to social vulnerability, health, and wellness. This participatory process recognizes individuals or groups as valued members and enables their contribution to resilient outcomes for themselves and the broader community. It integrates scientific knowledge, public perceptions, community concerns, local assets, and institutional capabilities.
3) **Implementation assistance** – The RAFT supports equity in implementation by connecting the dots and incorporating those services and programs that work with vulnerable populations into project implementation. This involves bringing service agencies (such as social service agencies, health departments, and transit agencies) into the coastal resilience network to address broader social and economic needs, such as food access, clean water, transportation, and healthcare. The implementation process emphasizes establishing and supporting local implementation teams to work on resilience projects, fostering long-term relationships and enabling adaptation to community needs.

Table 2.1 Examples of equity-related questions asked as part of The RAFT qualitative assessment

• Given your organization's work in this community, what people, neighborhoods, and/or populations are identified by your organization as historically excluded and underserved? • How does your organization serve these populations? What underserved people are not being served by you or other groups? • What strengths, needs, and opportunities in your community's resilience have been exposed, and for whom? (These could range from planning and leadership to infrastructure and health, food, shelter, and wellness. They could be exposed by a pandemic like COVID-19 and/or climate threats, such as storms, flooding, groundwater intrusion into septic tanks, saltwater intrusion into wells, or an extreme heat event that might impact access to critical services).

Developing The RAFT approach to equitable resilience

We have continued to learn how to incorporate and address equity in resilience planning and implementation throughout the development of The RAFT and through its application in different communities over the last decade. By continuously adapting our processes and tools based on lessons learned from communities, we strive to develop and apply a community-driven resilience planning and implementation process that fosters more equitable access, participation, and outcomes. The RAFT has continually evolved to integrate lessons learned and better respond to community needs.

During the initial development of The RAFT, an advisory committee composed of over 20 subject matter experts from government, nonprofit, and academic sectors provided essential advice and feedback on the content of the resilience scorecard and the broader process for engaging localities. This committee's input contributed to a holistic approach to resilience in The RAFT Scorecard, addressing a wide range of factors, many of which are connected to disparities and inequities, such as leadership, communication, community engagement, health, and well-being. Additionally, we consulted a group of social equity experts, who advised against creating a separate section for equity in The RAFT Scorecard. They recommended instead that it be integrated throughout every section. This led to modifications in scorecard components to include consideration of vulnerable populations, social equity, and health and wellness. Scoring metrics now assess the extent to which historically underserved and excluded groups are considered in existing practices, including risk assessments for socially vulnerable populations and public engagement in hazard mitigation planning targeted at these communities.

The advisory committee also helped develop key metrics for identifying, mapping, informing, and engaging socially vulnerable populations, and for working with trusted communication messengers and networks. This feedback underscored the importance of terminology, leading us to use "social equity" rather than "environmental justice" in some communities, as it resonated more broadly.

We also incorporated lessons learned about equity into process improvements. The process for selecting projects for the Resilience Action Checklist was refined to encourage the inclusion of a mix of short-, medium-, and long-term strategic impacts, moving beyond simply projects that are doable in one year to those that the

Implementation Team can initiate and make substantial progress on in one year. Projects undertaken during the implementation phase have included efforts to increase communities' awareness of the need to identify and prepare their socially vulnerable populations for coastal and other hazards, such as mapping critical infrastructure serving them, creating asset maps of services, and developing accessible emergency information.

Insights from research by students on the impacts of sustainability and resilience report cards, conducted as part of The RAFT development, revealed that these assessment tools were insufficient for catalyzing change, primarily due to reported resource constraints at the local government level for translating the report cards into action. This led The RAFT team to include a year-long implementation phase to help translate the scorecard findings into specific resilience-building actions. As part of this implementation phase, drawing on best practices in community engagement and collective impact theory, which emphasize co-production and partnership approaches, The RAFT process included the creation of community-based Implementation Teams to co-steward the implementation of the community's resilience priorities. The year-long implementation phase provides ongoing assistance to localities and tribal nations in implementing their Resilience Action Checklist items, with a specific focus on connecting services and programs that work with vulnerable populations.

Early experiences emphasized the importance of broad and durable community involvement, leading to modifications in the process, such as intentionally broadening the invitation list for community workshops to include a wider range of interests and members of socially vulnerable groups. Early experiences also revealed that local governments might unintentionally limit community involvement in The RAFT by only suggesting familiar members of the public to participate rather than the full range of vulnerable populations, calling for efforts to purposefully enhance Implementation Teams by seeking out community volunteers. The RAFT team adds capacity to do that by providing administrative support for these meetings, supplementing the limited staff and resources of many rural localities.

The RAFT also supported the development of a self-standing Regional Resilience Equity Workgroup for the Middle Peninsula and Northern Neck regions of Virginia, in recognition of the need, expressed by service organizations, for regional coordination and a support structure to sustain equitable resilience over the long term. With funding from a community foundation, The RAFT also introduced a mini and micro grant funding program to directly support community organizations' efforts to address the priority risks and needs of vulnerable populations through subgroups focused on economic development, healthcare, food security, affordable housing, transit, and youth programs. To reduce the application burden and make the funding opportunity more accessible and equitable for under-resourced programs, organizations, or communities, we used a one-page application form and a simplified performance reporting process. Examples of resilience projects supported through the mini and micro grant funding are provided in Chapter 4.

Three complementary forms of justice underpin The RAFT process approach:[12]

1) **Procedural justice** – This emphasizes equitable participation in decision-making processes, ensuring that communities are involved in decisions affecting them, including the development of plans and general governance. It is about *how* and *by whom* decisions are made.
2) **Distributive justice** – This focuses on the fair allocation of the benefits, costs, and adverse effects of resilience, ensuring equitable access to infrastructure, amenities, services, and economic opportunities for all, especially the most disadvantaged. It addresses *who gets what, why, and how much.*
3) **Recognitional justice** – This requires acknowledging and respecting different identities (e.g., race, gender, class, age) and understanding how historical injustices shape vulnerability, access to resources, and capacity to participate.[13] It ensures respect for different individuals and groups within a community, fostering inclusive and collaborative approaches.

Box 2.1 Access

Access refers to the ability of all community members to obtain necessary information, services, and opportunities.

- **Information accessibility**: This includes developing easily accessible emergency information for residents, providing translation services for non-English speakers as needed, and ensuring critical service information is channeled to vulnerable communities through trusted messengers and networks. For example, Spanish-speaking interpreters and/or translated materials can increase program accessibility.

- **Service accessibility**: Ensuring physical access to services like grocery stores, medical appointments, and shelters is crucial, especially in rural areas where transportation can be a barrier. Improving accessibility for people with disabilities in public spaces and programs, such as community gardens, is also important.
- **Resource accessibility**: Addressing disparities in access to financial resources and technical support is vital, particularly for under-resourced rural communities. The RAFT, for instance, provides administrative support to localities with small staff and limited resources to facilitate resilience planning and builds a supportive network of state agencies, nonprofits, and university partners.

The RAFT embodies several key principles and features of equitable resilience planning.

1) **Community-based and inclusive approach** – The RAFT emphasizes community involvement with diverse actors in the community, bringing their knowledge, risk perceptions, expertise, and resources to inform decision-making. We aim to build long-term relationships and sustain resilience efforts, offering support beyond the initial process and helping localities and tribal nations build their capacity. It supports resilience planning that often directly impacts individuals, households, and businesses, while officially engaging with localities and recognizing regional interconnectedness. Actively inviting and involving members of socially vulnerable groups, as well as service agencies and community organizations that work with them, is critical. This helps to overcome the disconnect that can arise between what the government perceives as community needs and what communities actually want, which can lead to distrust. When local governments constrained community involvement, The RAFT broadened its approach to include more diverse interests. It intentionally broadens the invitation list

for participation in community workshops and Implementation Teams to include members of socially vulnerable groups. It is also important to note that while this is one of our "north stars," it is not easy to achieve, and we continue to strive to learn how to facilitate a process where the voices of the most vulnerable are represented.

2) **Systems approach to address interconnected challenges** – Equitable resilience planning, as facilitated by The RAFT, involves a systems approach that acknowledges the interdependencies between acute shocks (e.g., coastal hazard events) and chronic stressors (e.g., poverty, aging infrastructure, recurrent flooding, persistent structural inequalities like racial injustice and classism, and uneven access to resources). This expanded approach to resilience ensures explicit consideration of social, cultural, economic, political, and physical components that contribute to inequities and disparities. The RAFT explicitly connects resilience to broader concerns about social and economic resilience, including physical and mental health, access to care, jobs, and financial resources. It recognizes that addressing resilience equitably requires meeting the needs of vulnerable populations and involves connecting service agencies (dealing with issues such as opioid addiction, aging, transportation, and mobility) into the coastal resilience network. While acknowledging the importance of infrastructure, The RAFT recognizes that resilience extends to community well-being, including health and social equity concerns of residents. It moves beyond a sole focus on physical infrastructure projects to consider less expensive, non-structural options that can significantly strengthen coastal resilience through adaptation. For example, we worked with coastal communities to develop and disseminate emergency communication magnets that include emergency contact information, an emergency checklist, and media stations and websites for alerts (see Figure 2.2). These refrigerator magnets were designed to help residents, both owners and renters of property, quickly access critical information during disasters.
3) **Multifunctionality and integration** – The RAFT supports planning that is multifunctional, drawing on different planning instruments and integrating elements like disaster risk reduction, infrastructure improvements, historic preservation, building codes, zoning ordinances, and capital planning. It emphasizes an integrative approach to overcome policy silos and avoid maladaptive

outcomes, especially those stemming from isolated planning processes that perpetuate inequities.

4) **Academic objectivity and neutral facilitation** – The RAFT team brings academic objectivity to the resilience assessment, prioritization, and implementation phases. This reduces perceived political influence and the pressures of the political cycle, which can create short-term planning horizons that hinder long-term resilience efforts. The RAFT's academic objectivity and neutral facilitation allow for more authentic consideration of equity and a critical examination of social vulnerability and disparities. We act as a liaison between the universities and the community, translating research into action while incorporating local knowledge.
5) **Prioritization of vulnerable populations** – The RAFT process emphasizes vulnerable populations as a focus of action for building resilience, prompting communities to brainstorm ideas that address equity issues unique to them. Scoring metrics in The RAFT Scorecard explicitly account for the needs and disparities faced by socially vulnerable populations, as well as the extent to which historically underserved groups are considered. The RAFT seeks to help communities identify and prepare their socially vulnerable populations for coastal hazards and has supported efforts like mapping critical infrastructure that serves them (e.g., healthcare, fire and rescue services, schools, libraries, public transit, food pantries).
6) **Language, framing, and trust-building** – The RAFT recognizes the importance of terminology, finding that the term "social equity" may resonate more than "environmental justice" in some rural contexts, and emphasizing that terms used must resonate with the communities. In one region, for example, the term equity did not resonate. Instead, the community chose the word "livability" to express the importance of including everyone in ensuring the community's overall well-being. Terms often used in academic and professional circles, such as resilience, sustainability, and climate impacts, should be explored with the community to learn whether they would resonate and help bring the community together. Finding the right language through sustained engagement and broad community input helps build relationships and trust.

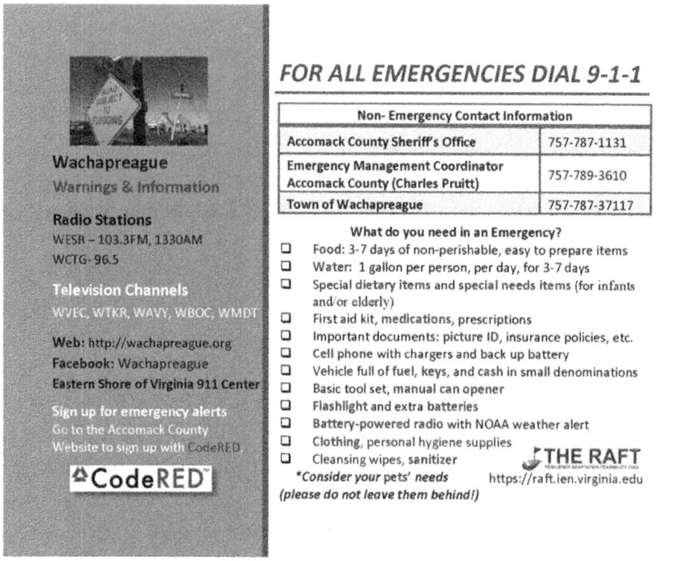

Figure 2.2 Emergency information magnet.
Source: Created by Michelle Covi. Used with permission.

Box 2.2 Emphasizing equity by connecting resilience to broader social and economic needs

The RAFT recognizes that communities have broader social and economic concerns that are intertwined with coastal resilience. The following are three examples that show how resilience projects can enhance equity by addressing social and economic aspects of resilience:

- **Holistic resilience priorities.** In Petersburg, a coastal-adjacent city, The RAFT process revealed priorities such as increasing access to parks and green spaces for urban heat relief, expanding mobility options for low-income residents to access jobs, and diversifying the city's economic base to withstand stress. The City of Portsmouth – one of The RAFT pilot communities – also provides a good example of a locality that took a holistic approach to establishing its resilience priorities. The City explicitly incorporated equity into its resilient thinking within its comprehensive plan, advocating for environmental and social justice in land-use decisions, such as denying a car wash application next to a new public housing project due to noise and aesthetic concerns. This approach reflects a shift towards balancing industry needs with community well-being. Portsmouth's comprehensive plan addresses economic diversification, living wages, and the incentivization of women- and minority-owned businesses as part of a resilience strategy.

- **Food access and security.** In two rural coastal communities (Essex County and King and Queen County), community members identified food access as a resilience priority, especially for the elderly and lower-income residents. The RAFT connected the Rappahannock Tribe leaders involved in heritage gardening with African American faith leaders and activists, as well as technical experts from the Virginia Cooperative Extension, to collaboratively develop a food security plan that involved community gardens.
- **Addressing septic system issues**. In the Middle Peninsula and Northern Neck regions of coastal Virginia, the issue of septic system inundation in underserved areas was highlighted as a significant resilience, environmental health, and public health issue impacting the local economy. During The RAFT application in these regions, we conducted an informational workshop to educate community leaders and residents about inundation risks and impacts on their septic systems, and we produced information sheets for the communities on sea level rise and flooding impacts on septic systems, wells, and roads.

Lessons learned for practice

1) **Integrate equity as a foundational principle** – Define and apply equity broadly and address social vulnerability throughout resilience efforts. Equitable resilience goes beyond mere equality (meaning equal allocation for all) to consider fairness and justice

based on existing needs and assets. Resilience efforts should recognize that hazards disproportionately impact low-income populations, racial minorities, and other vulnerable groups (such as women, children, older people, people with disabilities, and those facing financial instability or social isolation). Resilience planning and implementation must consider how past inequalities contribute to present-day uneven flood risk. It is vital to identify vulnerable populations (such as low-income populations, racial minorities, older people, people with disabilities, and those with medical fragility) who disproportionately bear the brunt of hazards and lack resources to cope. This can be done as part of a resilience assessment (see Chapter 3), by analyzing U.S. Census data, or by talking to community and faith-based leaders or locality staff.

2) **Integrate social and economic viability** – It is crucial to recognize that for many communities, particularly those that are under-resourced, economic and social resilience are often more urgent priorities than coastal hazards. Therefore, resilience planning must broaden the conversation to frame coastal hazards within the larger context of overall community resilience, connecting the dots between resilience to hazards and broader social and economic needs. This also means that social and economic dimensions must be explicitly included in resilience assessments and actions. Connect coastal resilience to broader social and economic resilience by explicitly linking coastal hazard preparedness with issues like food access, transportation, healthcare, and support for vulnerable populations. This means bringing social service agencies into the coastal resilience network.
3) **Assess systemic, long-term issues that challenge equitable resilience** – While immediate needs are important, a focus on equity requires addressing systemic issues that contribute to long-term vulnerabilities, such as poverty, lack of jobs, and limited access to financial resources. This can involve transforming practices that reinforce inequalities and ensuring resilience efforts lead to positive change for all.
4) **Address services that are critical for vulnerable populations** – Once critical services for vulnerable groups are identified, such as healthcare, food pantries, libraries, or schools, it is important to then identify and address specific barriers to accessing these services, such as transportation or internet access.

5) **Design broad and durable community-based and inclusive engagement** – Resilience planning with an equity focus requires broad and durable community involvement and representation of diverse interests from across the community, including non-profits, community organizations, and faith leaders, not just local government staff. This requires concerted effort, time, and resources to build relationships and trust. This requires moving beyond traditional public meetings that often exclude vulnerable populations due to imbalanced power dynamics, inconvenient locations, or culturally inappropriate agendas. The RAFT intentionally broadened invitation lists for workshops and implementation teams to include socially vulnerable groups and community volunteers. With funding secured from community foundations, honoraria were offered to individuals without organizational affiliations to compensate them for their time.
6) **Utilize a co-production approach** – Create a method or engagement process where community stakeholders are helping to inform and design the engagement process itself, including the goals and priorities. A co-production process can also bring together a wide range of stakeholders to collaboratively develop resilience actions. This encourages social learning, where diverse stakeholders share experiences, knowledge, and perspectives through dialogue and collaborative activities. This interaction fosters mutual understanding, trust, and collective problem-solving, enabling participants to adapt and innovate together. By reflecting and iterating together, communities build a shared vision and stronger resilience strategies. This approach integrates scientific knowledge, public perceptions, community concerns, and institutional capabilities to spur action and solutions.
7) **Employ skilled facilitation** – Skilled, impartial facilitators are essential to guide discussions, balance diverse views, and ensure all voices are heard without being swayed by political agendas. The facilitators' role is to guide conversations without imposing personal or political biases, manage conflicts constructively, and create a safe space where all participants, regardless of power or status, feel heard and valued. This approach helps prevent dominant voices from overshadowing marginalized perspectives, fostering trust and genuine collaboration in co-production processes.
8) **Work at the community's own pace** – Facilitators often talk about working at the speed of trust. There may be tensions

between parts of a community, between organizations, or even within the local government, so working together may need to proceed slowly. Use their language to help establish trust and buy-in. Allowing communities to come to their own conclusions by sharing stories and data, rather than imposing information, fosters engagement and belief in the process and resulting outcomes.

9) **Start with local leadership** – Local leadership is critical, as committed local champions and advocates can drive progress and sustain discussions beyond The RAFT year of implementation, even when political will or capacity is limited.
10) **Target resources, actions, and decisions** – Equitable resilience calls for identifying the most vulnerable and their needs, then targeting actions and resources to address those needs meaningfully, moving beyond just a paper exercise. This includes implementing specific projects that support vulnerable populations, such as mapping critical infrastructure that serves them and creating asset maps of services and resources. Dedicate specific actions and resources to redress disproportionate impacts on vulnerable communities, such as by prioritizing resource investments based on satisfying human needs, not just economic return on investment (ROI).

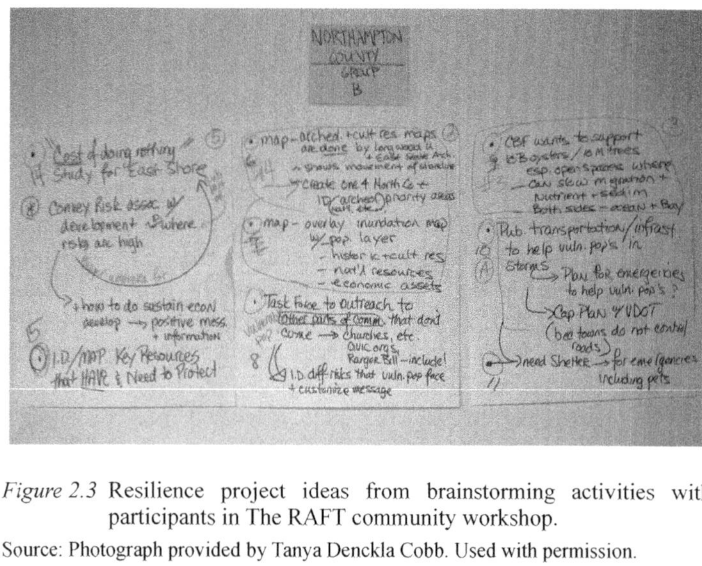

Figure 2.3 Resilience project ideas from brainstorming activities with participants in The RAFT community workshop.

Source: Photograph provided by Tanya Denckla Cobb. Used with permission.

11) **Recognize funding as an important priority** – Under-resourced communities may lack the capacity (staff, technical expertise, financial resources) to develop plans or apply for grants for resilience projects. Funding mechanisms, like mini and micro grants, can (as discussed in Chapter 4) directly support community organizations in implementing projects. Supporting a community in gaining access to such funding can be a transformational accomplishment.

Box 2.3 Examples of objective facilitation tools used by The RAFT team that can be used for community meetings

1. **Set clear ground rules** – Begin by establishing norms for respectful dialogue, equal speaking time, and avoiding dominance by any group.
2. **Active listening and summarizing** – Reflect back what participants say without judgment and summarize key points to show understanding.
3. **Allow anonymous input** – Use sticky notes, digital polls, or anonymous comment boxes so quieter voices can contribute safely.
4. **Balance power dynamics** – Break into small group discussions to reduce intimidation and allow marginalized voices to emerge.
5. **Use neutral language and framing** – Avoid loaded terms or political framing, present issues factually, and encourage evidence-based discussion.
6. **Use visual facilitation** – Use charts or diagrams to capture ideas objectively, so all contributions are visible and valued.

Notes

1 Manyena, S. B., O'Brien, G., O'Keefe, P., & Rose, J. (2011). Disaster resilience: A bounce back or bounce forward ability? *Local Environment, 16*(5), 417–424.
2 Rothstein, R. (2017). *The Color of Law: A Forgotten History of How our Government Segregated America.* Liveright Publishing.

3 Rivera, D. Z., Jenkins, B., & Randolph, R. (2022). Procedural vulnerability and its effects on equitable post-disaster recovery in low-income communities. *Journal of the American Planning Association*, *88*(2), 220–231.
4 Shokry, G., Anguelovski, I., Connolly, J. J. T., Maroko, A., & Pearsall, H. (2022). "They didn't see it coming": Green resilience planning and vulnerability to future climate gentrification. *Housing Policy Debate*, *32*(1), 211–245.
5 Wilson, B., Tate, E., & Emrich, C. T. (2021). Flood recovery outcomes and disaster assistance barriers for vulnerable populations. *Frontiers in Water*, *3*, 752307.
6 Rivera, D. Z., Jenkins, B., & Randolph, R. (2022). Procedural vulnerability and its effects on equitable post-disaster recovery in low-income communities. *Journal of the American Planning Association*, *88*(2), 220–231.
7 Flores, A., Collins, T., Grineski, S., Amodeo, M., Porter, J., Sampson, C., & Wing, O. (2025). Federally-overlooked flood risk inequities in the conterminous United States. *Scientific Reports*, *15*(1), 10678.
8 Knighton, J., Buchanan, B., Guzman, C., Elliott, R., White, E., & Rahm, B. (2020). Predicting flood insurance claims with hydrologic and socioeconomic demographics via machine learning: Exploring the roles of topography, minority populations, and political dissimilarity. *Journal of Environmental Management*, *272*, 111051.
9 Emrich, C. T., Tate, E., Larson, S. E., & Zhou, Y. (2020). Measuring social equity in flood recovery funding. *Environmental Hazards*, *19*(3), 228–250.
10 Emrich, C. T., Aksha, S. K., & Zhou, Y. (2022). Assessing distributive inequities in FEMA's Disaster Recovery Assistance Fund allocation. *International Journal of Disaster Risk Reduction*, *74*, 102855.
11 Anguelovski, I., Shi, L., Chu, E., Gallagher, D., Goh, K., Lamb, Z., Reeve, K., & Teicher, H. (2016). Equity impacts of urban land use planning for climate adaptation: Critical perspectives from the global north and south. *Journal of Planning Education and Research*, *36*(3), 333–348.
12 Yusuf, J.-E. W., Denckla Cobb, T., Andrews, E., Gladfelter, S., & Montrose, G. (2022). The Resilience Adaptation Feasibility Tool (RAFT) as an approach for incorporating equity into coastal resilience planning and project implementation. *Shore & Beach*, *90*(4), 53–63.
13 Matin, N., Forrester, J., & Ensor, J. (2018). What is equitable resilience? *World Development*, *109*, 197–205.

3 Measuring what matters

Using quantitative and qualitative resilience assessments tailored to community needs

Introduction

Resilience planning begins with an assessment of a community's resilience to identify what a community does well and what it still needs to do to become and/or remain resilient to natural hazards and climate impacts, as well as to chronic and acute stressors. To be comprehensive and useful, this assessment must be tailored to the community's needs and include both quantitative and qualitative components.

The Resilience Adaptation Feasibility Tool (RAFT) team has learned that having an objective assessment that relies on verifiable processes and evidence, and is conducted by an impartial party, is important for launching resilience planning efforts. Who conducts the assessment, and how, are as important as what is assessed. This chapter presents two resilience scorecards that provide a starting point for assessing a community's resilience quantitatively and discusses how these quantitative assessments define and measure resilience broadly while incorporating equity. In the case of The RAFT process, the impartial party is our three-university collaborative that brings the expertise and practical skills of faculty, practitioners, and graduate students in conducting the assessment. While universities can be seen as extractive by communities, and historically often have been, one strength of The RAFT process is that academics are also often trusted by both local governments and the wider communities that surround them as being impartial or not having deep vested interests in local politics in the same way that an advocacy group or other locally embedded group may have. We recognize that, as a practitioner reading this, you may be part of a nonprofit advocacy organization, service provider, or local government, with your own

DOI: 10.4324/9781003596233-3

relational politics in the communities where you plan to do resilience work. This does not mean that an assessment is not possible at all, but it is important to acknowledge that *who* conducts the assessment matters; and if there are underlying politics and historical baggage that compromise community trust, it may be necessary to collaborate with a third party evaluator or mediator, or invest in a much more robust community engagement process from the start that builds trust in where the data is coming from and whose perspectives are included in the assessment.

Resilience extends beyond simply numbers on a quantitative scorecard, however. Particularly if there are community trust challenges, incorporating the perspectives and resilience concerns of a diverse cross-section of the community in a qualitative assessment will be critical to build community buy-in and trust in the assessment data overall, especially as you proceed into implementation. Our experience implementing The RAFT in coastal communities in Virginia identified that resilience to hazards or climate impacts often takes a back seat to more pressing chronic and acute economic and social stressors. Conducting qualitative assessments of resilience helps communities prioritize resilience by helping them understand it beyond the numbers of a scorecard and by placing it within a broader context of the community's economic, social, and environmental issues. Additionally, when community members and service providers in a region see their own urgent resilience challenges elevated as shared priorities for investment, they are much more likely to participate in a broader resilience implementation process and invest their own energy and trust in other resilience efforts that may not be their top priorities. This chapter offers such a qualitative approach for understanding resilience from a community perspective by using listening sessions with community, business, educational, nonprofit, and governmental stakeholders, drawing on examples from qualitative assessments conducted for coastal communities in Virginia.

Any community resilience assessment effort must be iterative to stay relevant, accurate, and effective through time. The RAFT process has been an iterative one for over a decade of development and implementation, catering to the disparate needs of different regions and reflecting changing environmental, social, and economic conditions. We next describe The RAFT assessment strategies – quantitative and qualitative – along with some key insights into their development and application.

The RAFT quantitative assessment: Evolving to address changing conditions and varying community needs

Assessment of resilience must be an iterative process if it is to remain relevant and effective. Thus, The RAFT Scorecard has evolved to address changing climate impacts and emerging community challenges, and to incorporate research findings and stakeholder input. Through that iterative process, we have learned that having an objective assessment that relies on verifiable processes and evidence, and is conducted by a party perceived as neutral by participating localities, is important for launching effective resilience planning efforts. Local governments sometimes distrust or have difficult relationships with community and nonprofit organizations that engage in advocacy work or proactive litigation, so neither party trusts the other to conduct an objective assessment. In the case of The RAFT, the coalition of three state universities that comprise The RAFT team serves as an impartial, objective party. However, a well-trusted nonprofit or community organization that has been active, effective, and supportive in a community long enough to build relationships may also be positioned to conduct an assessment that is received as objective and accurate, particularly if the group conducting the assessment invests in building trust in the process.

In general, universities, nonprofit boundary-spanning organizations, regional entities such as planning organizations, and extension services can play a key role as partners in conducting such assessments by providing information on existing efforts and policy gaps in the community. However, the leadership of recognized neutral entities is key to a successful quantitative assessment conducted in partnership with locality staff and others. It is also important that the assessing entity or entities have the necessary skill sets and expertise to develop and administer the assessment tool. For example, The RAFT team expanded from its original partners at the University of Virginia and William & Mary Law School to include Virginia Sea Grant extension staff at Old Dominion University, adding to the team's expertise so that it now included a broader array of expertise in science as well as planning, mediation and facilitation, law, and community engagement.

In addition, it's important to recognize that maintaining the accuracy of the assessment tool and conducting the scoring is time- and resource-intensive. Close attention must be paid to changing laws, regulations, and policies at different levels of government, and it requires extensive time to pore through local ordinances, plans, and policies to determine whether they incorporate specific resilience elements and to then meet with local government officials to ground-truth the scoring. That grassroots analysis, using an accurate and time-tested assessment tool, is a major factor in The RAFT's success. Finding funding for the scoring and having access to knowledgeable individuals with the expertise to conduct the analysis, preferably with already-established relationships at the regional or local level, so they are familiar with the political and economic setting and legal constraints, are significant challenges when creating an effective resilience assessment process.

In this chapter, we discuss both The RAFT Scorecard and the PREP (Planning for Resilience: Evaluation and Prioritization) Tool, a local government self-assessment tool developed by The RAFT team, as examples of methods for quantitatively assessing a community's resilience to a range of hazards, climate impacts, and other factors that affect the community's well-being. Practitioners are welcome to try both tools and to adapt them as needed to best suit their circumstances. Other assessment tools are also available to communities seeking to increase their resilience. Importantly, each community must determine what approach is best for them, and we hope that the lessons we share here from developing and using The RAFT Scorecard will help you in making that determination.

The RAFT Scorecard and scoring process

A quantitative assessment of a community's resilience needs to be comprehensive, covering multiple dimensions. To achieve this, The RAFT uses a broad definition of resilience. Initially, this definition was: being able to anticipate, adapt, endure, and thrive in the face of change, uncertainty, and adversity. A decade later, we define resilience as the ability to respond to acute and chronic hazards and stressors while remaining economically and socially viable. The RAFT focuses on community resilience, improving a community's ability to bounce back from hazard events such as hurricanes and extreme heat, and deal with chronic issues such as flooding. As such, The RAFT's definition takes a comprehensive approach by including environmental, economic, and social resilience. All three are vital for a community to thrive. While every resilience effort may formulate its own locally grounded definition of resilience, we have found that including multiple dimensions of resilience is essential to enable a broader, more diverse group of organizations and service providers to see themselves as having a role to play and resources to contribute to implementation.

Box 3.1 Changing definition of resilience

The RAFT definition of resilience has changed over time. In the early years of The RAFT development and application, we defined resilience as the ability to anticipate, adapt, endure, and thrive in the face of change, uncertainty, and adversity. Over the years, as we have worked with more communities and seen them endure the COVID-19 pandemic and economic downturns, we have expanded the definition of resilience we use. Currently, the definition of resilience that informs our approach to working with communities is to increase their ability to respond to acute and chronic hazards and stressors while also strengthening their economic and social viability.

It is important to base the assessment of resilience on a broad definition of resilience, one that encompasses more than simply emergency response and planning for impacts, to include social and physical infrastructure, natural- and nature-based features, and health and well-being. Reflecting that broad approach, The RAFT Scorecard (our quantitative assessment tool) considers resilience efforts in five categories (see Figure 3.1):

1) **Policy, leadership, and collaboration** – Measures policy and legislation in place for resilience and includes coordination and collaboration between various levels of government, and how accessible and open resilience-related information is to the residents and other members of the community.
2) **Risk assessment and emergency management** – Examines whether risk assessments have been conducted to address relevant hazards such as flood, storm, wildfire, drought, and extreme temperature, identified vulnerable populations and their needs, and developed plans for disaster preparedness and response, including a Hazard Mitigation Plan.

1. Policy, leadership, and collaboration

1.1 Leadership and planning for resilience
1.2 Leadership and responding to emergency
1.3 Collaboration with state and regional agencies
1.4 Adaptive management
1.5 The NFIP's Community Rating System

2. Risk assessment and emergency management

2.1 Flood exposure and vulnerability assessment
2.2 Risk assessment for vulnerable populations
2.3 Business and economic risk assessment
2.4 Hazard mitigation
2.5 Resident emergency preparedness

3. Infrastructure resilience

3.1 Stormwater infrastructure
3.2 Critical transportation infrastructure
3.3 Water supply and wastewater management services
3.4 Utilities infrastructure
3.5 Critical infrastructure for emergency services
3.6 Flood control infrastructure
3.7 Natural and nature-based features

4. Planning for resilience

4.1 Budget, funding and state and federal assistance
4.2 Hazard resilience in comprehensive plan
4.3 Land use ordinances
4.4 Incentives for hazard resilience
4.5 Natural resilience preservation

5. Community engagement, health, and well-being

5.1 Community involvement in resilience planning
5.2 Providing hazard resilience information to the community
5.3 Community leadership and volunteer networks for resilience
5.4 Resilient systems to provide food, health, and medicine
5.5 Physical and mental health in community resilience

Figure 3.1 Components of The RAFT Scorecard.
Source: Created by authors.

3) **Infrastructure resilience** – Assesses whether there are identified methods and plans for protecting critical infrastructure from climate impacts and hazards, including the use of natural or nature-based features.
4) **Planning for resilience** – Assesses the comprehensive plan and zoning code for resilience, how a locality or tribe is using incentives to promote resilience in building and development, how policies protect ecosystems, how they use green infrastructure to improve resilience, and how much resilience has been incorporated into planning.
5) **Community engagement, health, and well-being** – Assesses how the locality or tribal government and staff engage with residents in planning for flood, storm, wildfire, drought, or extreme temperature hazards, including social equity considerations, and examines the locality or tribe's attention to issues of health and wellness during and after hazards

The fifth scorecard category, Community Engagement, Health and Well-being, is often the category in which localities achieve the lowest score, and sometimes surprises people that such programs and policies are included in a resilience scorecard. Yet a community's social and economic health must be considered in such an assessment, alongside environmental resilience, because these factors significantly determine whether and how well a community prepares for and responds to chronic stressors and hazardous events.

In The RAFT model, scoring is generally conducted by students under the supervision of a professor at one of the three partner universities, in consultation with The RAFT team as needed. An alternative model is used when The RAFT collaborates with sovereign tribal nations to protect the tribes' privacy and data sovereignty. In this case, a Tribal Liaison, appointed by the Chief from within the tribe, conducts the scoring and coordinates with the Tribal Council, elders, and tribal members (more on this in Chapter 6). Overall, in our scoring efforts, we found that a robust assessment requires a team effort, with input and ground truthing from locality staff across a wide range of functions and areas of expertise, which would also be true of an assessment conducted by a non-academic team of practitioners. By not assigning the scoring to one person, The RAFT assessment process adds objectivity, capacity, expertise, and resources, making the assessment more achievable and credible. In the case of the Tribal Liaison, the burden of time and labor associated with scoring is offset by a stipend provided to this individual, and the liaison still collaborates actively with a team of staff and tribal leaders to complete the assessment.

As a practitioner reading this guide, you could choose to use a similar approach with similar scoring categories as those used in The RAFT Scorecard, since those should translate fairly well to other regions facing similar hazards; or, you could choose to use The RAFT Scorecard and the PREP Tool as starting points and adapt them and the scoring process as needed for different circumstances. For example, if you are not part of an academic team and do not have access to students to assist with scoring, you could partner with others in the region who are familiar with local government processes to conduct the scoring. Each situation will require the determination of relevant resilience challenges and appropriate responses to be assessed by the scoring, yet the concepts and principles concerning resilience and centering equity can remain the same, no matter who conducts the scoring or what hazards are addressed.

The development of The RAFT Scorecard

The concept of The RAFT sprang from a meeting in 2015 about sea level rise and climate change with local governments in coastal Virginia (see Chapter 2 for the history and development of The RAFT). Some of the participating representatives of coastal localities noted that, due to the political situation, it was difficult to initiate discussions about planning for climate change impacts in their communities, and they wanted assistance in initiating those conversations.

To determine how best to respond to the localities' concerns, a team of graduate students in the Urban and Environmental Planning program at the University of Virginia surveyed more than 30 existing resilience and sustainability scorecards and report cards, analyzing specific metrics and scoring approaches. Five were selected for further analysis. This subsequent analysis included interviews with scorecard developers and users to identify lessons learned and recommendations for creating and using such tools. The team then convened an advisory committee to provide guidance on developing scorecard content to ensure an effective assessment of a locality's resilience. The advisory committee included over 20 subject matter experts from state and local government, universities, and environmental nonprofits.

Figure 3.2 A resident reading The RAFT Scorecard.
Source: Photograph provided by Virginia Sea Grant. Used with permission.

Originally, The RAFT team considered developing two sets of scorecard indicators, one each for urban and rural localities. After an initial attempt to do that, we quickly found that the distinctions between the two types of communities were not simple or clear, as many localities have both urban and rural characteristics. With input from The RAFT advisory committee and after extensive discussion, we decided that the assessment needed to be broad enough to apply to both rural and urban communities.

The advisory committee contributed many recommendations about the scorecard and scoring process, including:

- The assessment tool should be a scorecard rather than a report card with an A to F grading scale, so that it would be seen as helpful rather than punitive and the focus would be more on needed resilience measures and flagging opportunities to improve resilience, rather than the grade received.
- The scorecard should have clearly described indicators, specific sources or methods for measuring each indicator, and a clear scale

for assigning numbers or scores so that the measurement is as transparent, consistent, and objective as possible.

- The scorecard indicators should be developed objectively by The RAFT team, not by communities, and should not vary from community to community, nor be subject to local political pressures. The scoring should be done objectively by The RAFT team as well; local governments should not self-score.
- As much as possible, subjectivity and the need for judgment calls in scoring should be eliminated, with each indicator being answerable by a clear yes/no. Metrics should, to the extent possible, be based on readily available information, as transparency and easy access to information are important contributors to community resilience.
- The RAFT Scorecard should overcome a problem identified in the analysis of other scorecards – that they increased awareness of resilience but did not lead to local action and changes that actually improved resilience – by also providing guidance for setting a community's resilience action priorities. Each indicator of The RAFT Scorecard should be translatable into specific resilience-enhancing action.

Box 3.2 The important role of an advisory committee

Creating an advisory committee composed of experts and practitioners from state, regional, and local government, academia, and nonprofit organizations who are experienced with resilience challenges and programs is exceptionally helpful when designing a resilience assessment process. Such a committee can provide the needed expertise, oversight, and neutrality, building community trust and adding credibility to the assessment process, even if it is led by a local government or nonprofit perceived as having a biased viewpoint.

For example, The RAFT team convened an advisory committee to provide guidance on the content of The RAFT Scorecard, the appropriate scorecard metrics, and the overall approach and process. The advisory committee included academics from multiple universities, representatives of state government agencies, nonprofit organizations, The RAFT pilot communities, and members of a focus group that discussed the metrics to be used in the scorecard. In addition, The RAFT Team convened a social equity focus group, with participants providing detailed insights, revisions, and additions to the scorecard to ensure it would be robust in addressing social equity in locality resilience efforts.

To assess the draft scorecard, we formed a small group composed of representatives of state government agencies and the planning district commissions (the equivalent of regional planning organizations) involved with coastal issues. The group provided feedback on the scorecard's objectivity and effectiveness in assessing locality-specific coastal resilience. It recommended a regional approach for The RAFT to enable comparisons between localities within the same region, foster regional coordination and discussion, and enable all Virginia coastal localities to participate as rapidly as possible. The group recommended working in partnership with the regional planning agency, and those partnerships have been key to The RAFT's success. Regional staff are familiar with local risks and challenges; they know their localities, and their localities can feel more comfortable participating in The RAFT knowing that regional planning commissions are involved. They can also introduce The RAFT team to local staff for assistance with finding materials during the scoring and vetting of results and can provide connections to local governing boards for presentation of the scorecard results.

We also conducted a focus group of social equity experts to discuss how to ensure equity was incorporated into the draft scorecard. The group advised that equity should be incorporated throughout all sections of the scorecard, not as a separate category, and provided detailed insights and recommended revisions to achieve that. The need to identify and engage socially vulnerable residents is identified throughout the scorecard. For example, subsection 4.2 of the scorecard (focused on hazard resiliency in comprehensive planning) notes that "The ideal comprehensive plan identifies equity and the need to identify and support socio-economically vulnerable populations as a priority for resilience." Subsection 5.5 of the scorecard focuses on incorporating social equity as an element of physical and mental health in community resilience.

Initially, the scorecard was tested across three geographically diverse pilot communities: a town, a city, and a county. Two of the localities scored themselves concurrently with The RAFT team, and provided very helpful input into the resilience indicators and scoring process. Representatives of the three pilot communities also served on The RAFT advisory committee ensuring that their experience with application of the scorecard was included in the committee's discussions.

As of 2026, The RAFT has been applied in three coastal regions and two coast-adjacent cities in Virginia and is currently supporting resilience planning and implementation in two localities in southwest Virginia and three localities in the Historic Triangle of Virginia. The scorecard has evolved to incorporate heat and more inland concerns, such as riverine flooding and wildfires. We have also modified The RAFT Scorecard and process to work with four Virginia Indian Tribes, who, as sovereign nations, conduct their own scoring with the support of The RAFT team members. This is important to note as it highlights how the foundational underpinning of The RAFT Scorecard categories allows for additions and revisions to account for specific community risks and contexts.

The PREP (Planning for Resilience: Evaluation and Prioritization) Tool

The RAFT's advisory committee recommended that the scoring be done by objective parties, rather than as a locality self-scoring exercise. However, in light of The RAFT's success with grassroots engagement and building resilience in communities, some state and local government representatives asked us to also develop a self-assessment tool for local governments, in hopes of more quickly expanding access to The RAFT across more communities. In response, we developed the PREP Tool for use by local governments.

The PREP Tool provides a structured three-step process to guide local governments through a self-assessment process that results in a resilience plan informed by risk level, existing resilience efforts, and best practices in resilience planning. The assessment component of the PREP Tool offers another example of a quantitative resilience assessment. Unlike The RAFT Scorecard, the PREP Tool is intentionally designed to be self-scoring, with locality staff evaluating their own plans, policies, and practices.

The PREP Tool is a self-scoring assessment designed to be completed by a team of locality staff across multiple departments or functions, rather than by a single staff member. Each section of the PREP Tool identifies the types of staff and stakeholders best positioned to complete it. The PREP Tool assesses resilience across planning, infrastructure, emergency management, social equity, economic resilience, and natural systems. Completing the assessment, therefore, requires input from planners, emergency managers, public works and engineering staff, social services providers, finance or budget staff, and others with subject-matter expertise. When completed by a single individual, the assessment risks being incomplete or overly influenced by one departmental perspective. When completed by a cross-functional team, the process itself becomes a mechanism for internal coordination and shared understanding of resilience challenges.

The PREP Tool uses a series of structured yes-or-no questions that local staff score themselves based on existing plans, policies, programs, and practices. Because the tool is self-scoring, its value lies less in producing a defensible external benchmark and more in supporting internal reflection, discussion, and prioritization. Disagreements

Figure 3.3 The PREP Tool.

Source: Created by authors.

among staff about how to answer a question often reveal gaps in communication, documentation, or coordination that are important resilience issues to address. The PREP tool recommends dividing sections among sub-teams and then reconvening to review results collectively. This step allows staff to compare findings across categories and identify common strengths and cross-cutting gaps or weaknesses.

This emphasis on team-based self-assessment distinguishes the PREP Tool from The RAFT Scorecard. While the scorecard relies on an impartial third party to ensure objectivity and comparability across communities, the PREP Tool prioritizes accessibility and internal capacity-building. It is intentionally designed for use by local governments themselves, recognizing both the resource constraints many communities face and the importance of internal buy-in for resilience planning.

Like The RAFT Scorecard, the PREP Tool approaches resilience comprehensively across multiple dimensions. The self-assessment is organized into seven categories that collectively reflect institutional, social, environmental, and infrastructural dimensions of resilience.

1) **Risk assessment** – Evaluates whether the locality has established a baseline understanding of current and future hazard risk. Example indicators include whether the locality has completed a flood exposure or vulnerability assessment within the past five years, whether climate change impacts such as sea level rise or increased precipitation are explicitly evaluated, and whether vulnerable assets and populations have been identified and mapped.
2) **Planning and land use** – Assesses the extent to which resilience has been incorporated into long-range planning and regulatory frameworks. Example indicators include whether the comprehensive plan acknowledges climate impacts, whether resilience projects are included in the capital improvement plan, and whether zoning or land use ordinances limit or discourage development in flood-prone areas or promote low-impact development practices.
3) **Social equity** – Focuses on identifying and addressing disproportionate impacts of hazards on vulnerable populations. Example indicators include whether the locality has identified socially or economically vulnerable communities, whether those communities are engaged directly and meaningfully in resilience planning, whether outreach materials are accessible (e.g., multilingual),

and whether projects are prioritized based on benefits to directly impacted or historically excluded populations.

4) **Economic resilience** – Examines the locality's ability to anticipate and respond to economic disruptions caused by hazards. Example indicators include whether the locality has assessed economic impacts from flooding or storms and whether businesses are engaged in preparedness planning.
5) **Emergency management** – Assesses preparedness, response capacity, and access to critical services. Example indicators include whether the locality has an up-to-date emergency operations plan, whether residents receive regular preparedness outreach, and whether shelters and evacuation options are accessible to low-resourced communities.
6) **Natural and nature-based features** – Evaluates how natural systems and green infrastructure are incorporated into resilience strategies. Example indicators include whether the locality has identified and mapped vulnerable natural resources, whether wetlands or marshes are used as protective features against flooding or erosion, and whether incentives or education programs encourage the use of nature-based solutions.
7) **Infrastructure** – Focuses on the resilience of stormwater systems, critical facilities, utilities, and transportation networks. Example indicators include whether stormwater infrastructure has been assessed for vulnerability to flooding, whether critical facilities located in flood-prone areas have adaptation plans, and whether transportation infrastructure vulnerabilities have been identified and addressed.

The PREP Tool is freely available for use in conducting a quantitative assessment of resilience. The indicators are directly applicable to localities subject to coastal hazards but can also be modified for use in other contexts. It is important to note that the use of the PREP Tool may achieve a quantitative assessment fairly quickly if there is adequate local staff to take on the scoring tasks, but it does not result in any qualitative assessment, which is a key yet time-consuming element of any community resilience effort. The PREP Tool itself encourages local government staff to also engage in efforts to obtain community input. While practitioners may have limited funding and time to conduct their resilience assessments and advance priority resilience projects in a particular community, investing in securing robust resources for a more time-intensive process will strengthen your overall process by building trust and community buy-in. Creating an advisory committee of experts to help oversee the process and provide input can also provide additional capacity and preserve trust in the assessment process.

The RAFT qualitative assessment

As comprehensive and detailed as The RAFT Scorecard is, scoring strengths and weaknesses is not enough. It gets public attention, particularly when the scores are presented to local elected officials at a public meeting, but the quantitative assessment may not be sufficient to really move the needle on resilience among community members. To increase input from local leaders and community buy-in, The RAFT team added a concurrent qualitative assessment, which, along with the quantitative assessment, feeds into the regional community workshop and the selection of prioritized resilience actions for the one-year implementation phase.

The qualitative assessment involves The RAFT team holding a series of listening sessions in the form of interviews and focus groups with community service providers and community organizations with the goal of understanding how community leaders view challenges, strengths, and opportunities for building resilience locally. The information gathered from this assessment is then combined with the findings from The RAFT Scorecard assessment to inform the next phase of the resilience planning process.

Development of The RAFT qualitative assessment

The COVID-19 pandemic required The RAFT team to shift to a completely virtual process for engaging localities in resilience planning. However, the pandemic also enabled a profound expansion in how we approached and defined resilience, making it much more inclusive of a range of sectors and issues that contribute to a holistic perspective on community well-being and resilience.

Recognizing the unprecedented nature of the COVID-19 pandemic and its relevance to resilience planning more broadly, members of The RAFT team at the Institute for Engagement & Negotiation at the University of Virginia received a COVID Rapid Response Grant from the University's Environmental Institute to conduct a study on how the COVID-19 pandemic was impacting critical services and the resilience of residents across the Northern Neck region of Virginia. Information was gathered through interviews and focus groups conducted virtually with over 60 representatives from more than 40 social service organizations across the region, and used to assess the impacts of COVID-19 on the delivery of critical services to vulnerable populations, with a focus on building long-term resilience to future disasters.

When The RAFT was next applied to support resilience planning and project implementation for localities of the Middle Peninsula region of Virginia, we used the interview and focus group model developed during the COVID-19 pandemic to consult on-the-ground stakeholders and conduct a qualitative resilience assessment at the initial stage of The RAFT process. In the Middle Peninsula, 65 individuals representing 60 different community service organizations and institutions were consulted in the listening sessions that informed the qualitative assessment. The organizations and entities represented included ten local and regional government agencies, 26 nonprofit or community organizations, 11 education or service organizations, seven state or federal agencies, five faith groups, and one Virginia Indian Tribe.

The qualitative assessment in the Middle Peninsula found that the region faced significant environmental, economic, and social challenges, but also possessed notable strengths. For example, environmental threats such as flooding, sea level rise, and failing septic systems posed risks to both coastal and inland communities, many of which lack emergency infrastructure and insurance coverage. Economic barriers included limited funding for service organizations, scarce employment opportunities, poor broadband access, and a shortage of affordable housing. Socially, underserved populations, including youth, seniors, tribal members, immigrants, and people with disabilities, struggled with limited access to healthcare, mental health services, childcare, and other essential resources.

Despite these challenges, the assessment also revealed strong social cohesion and community pride. Local organizations, nonprofits, and government agencies collaborated effectively, and residents supported one another in times of need. The region's natural environment, educational institutions, and cultural heritage were identified as valuable assets. Opportunities for strengthening resilience included improving disaster planning, expanding broadband and transportation access, increasing affordable housing, enhancing services for youth and older people, and fostering inclusive partnerships with historically excluded communities. These findings, based on the lived experiences of service providers and community leaders, offered a foundation for future planning and investment in resilience across the Middle Peninsula.

The results of this qualitative assessment deeply informed and strengthened the equity focus of our resilience planning and implementation process. Although time-intensive, the interviews and focus groups allowed for much more robust participation in The RAFT process from diverse community members and service providers and directly impacted the kinds of Resilience Action Checklist priorities that the localities ended up selecting to advance resilience. For example, through the qualitative assessment, we were also able to recruit a much more diverse group of stakeholders to engage in the project prioritization and implementation phases of the process. By inviting and holding space to hear the resilience challenges of diverse community members first, it created an inclusive starting place for those individuals and organizations to stay engaged in The RAFT process beyond the assessment phase. Many groups had never been involved in resilience initiatives or planning processes before, and many did not initially see their work as relevant to coastal or climate resilience. The listening sessions thus also educated participants about how their services related to resilience and created opportunities to bring fresh perspectives and new avenues for collaboration into resilience planning.

This equity-centered assessment also directly influenced the development and implementation of the Resilience Action Checklists for each locality. The community workshop that launched implementation was much more robust and diverse than past workshops. For example, 89 individuals participated in the virtual Middle Peninsula Regional Resilience Workshop held in January 2022, where they created their communities' Resilience Action Checklists, formed local Implementation Teams to work on the selected items for one year, and established their implementation priorities. At the time, this was The RAFT's most robust regional workshop in terms of total participants and diversity, including representatives from over 55 different institutions, including local and regional government, nonprofits and community organizations, education and service providers, state and federal agencies, faith groups, and Virginia

Indian Tribes, as well as individual residents. Additionally, many of the Resilience Action Checklist priorities were directly aligned with equity issues intersecting with resilience, as the diverse individuals and organizations who participated in the assessment phase were also at the table in the Resilience Action Workshop, where resilience priorities were identified and elevated for implementation support.

Essex County, for example, had equity integrated explicitly into all five items on its Resilience Action Checklist, with the priorities being:

- Improving community-engaged disaster preparedness.
- Increasing affordable housing and emergency shelter options.
- Expanding food access and food security through community gardens on public and private lands.
- Protecting and mitigating impacts on the availability of clean water.
- Improving access to vulnerable residents' homes before, during, and after storms.

In a more traditional planning process where only those individuals typically involved in resilience planning efforts (such as emergency managers, flood plain managers, zoning administrators) may have been present, several of the critical resilience needs on this list like affordable housing, food access and security, and clean drinking water protection, would not have likely become the focus of a resilience implementation process. This reveals the importance of including a broad spectrum of resilience stakeholders and their needs from the beginning in a resilience planning process, and genuinely involving them throughout implementation, if one hopes to build resilience in a comprehensive and integrated way. After all, the more diverse perspectives and talents on the team, the more possible it is to tackle the most wicked issues (or complex, intractable issues)[1] with both an intersectional perspective and creative solutions that will stick and last in a community. An important lesson from The RAFT process is that it takes significant time to conduct thorough outreach efforts to obtain grassroots community input. However, these initial efforts pay off because they result in a more representative qualitative assessment and broader attendance at the community workshop – both of which lead to the development of a more inclusive Resilience Action Checklist of projects and a more extensive Implementation Team to work on them.

Ultimately, the qualitative assessment of the COVID-19 pandemic's impact on the Northern Neck and a more general assessment of resilience barriers and opportunities in the Middle Peninsula had a significant long-term effect, reshaping our approach and process for working with localities and sovereign tribal nations, particularly during the resilience assessment phase. It also radically expanded the way in which we approach resilience, such that subsequent applications of The RAFT not only focused on more traditional areas of resilience like flooding and sea level rise, but also attainable housing, food security, access to mental and physical health services, and other underlying needs that build and maintain a community's resilience and overall well-being. These issues were included in future iterations of The RAFT, not through a top-down enforcement of equity-related themes, but rather by allowing them to emerge as priorities when diverse resilience stakeholders (such as librarians, school teachers, nonprofit leaders) beyond those typically included in resilience or disaster preparedness processes were sought out and actively invited to participate in resilience planning.

This model of conducting interviews and focus groups at the beginning of The RAFT process has been continued in applications of The RAFT since our work in Virginia's Middle Peninsula region and has become a critical piece of The RAFT assessment phase, which now includes both the quantitative scorecard assessment and qualitative interviews and focus groups held with community leaders and service providers to understand regional resilience assets, opportunities, and challenges. The RAFT's qualitative assessments, like the one in the Middle Peninsula described above, have varied somewhat by region but generally have involved four to five focus groups, 15 to 20 individual interviews, and a survey to capture additional perspectives. In past regional qualitative assessments, we have consulted between 44 and 65 individuals representing anywhere between 38 and 60 unique organizations and institutions serving a particular region.

While not every qualitative resilience assessment will look the same or model The RAFT's approach, it is important to allow for significant time and to seek resources to conduct a robust qualitative resilience assessment. Being inclusive in the qualitative assessment requires doing extensive research and stakeholder mapping to identify groups that have not previously been included in resilience planning, followed by proactive outreach to invite them to a listening session scheduled around their availability. In addition, decisions need to be made about the best format for listening sessions (virtual, in-person, or hybrid), depending on community members' needs and capacity.

Community foundations or other private funders with a local focus in the specific geographic area may be interested in supporting the qualitative assessment component of your resilience work. Alternatively, it may be possible to partner with a local university or community college to have faculty and students conduct the qualitative assessment pro bono as part of a class or independent study.

Our experience implementing The RAFT in coastal communities in Virginia identified that resilience to hazards or climate impacts often takes a back seat to more pressing economic and social issues. The listening sessions with community leaders, nonprofits, service providers, and local government staff provided an opportunity to elevate and center the other economic and social issues facing the community. Moreover, the qualitative assessment process creates connections across the community that feed into the regional workshop and resilience priority setting and then continue during the year of implementation and beyond The RAFT process. The value of these interviews and focus groups is not just in the qualitative assessment it produces that focuses on assets, opportunities, and challenges while connecting equity with resilience, but also that the process has resulted in more buy-in and a willingness among community leaders to engage in the project prioritization and implementation processes because the resilience priorities resonate with their experiences and vision of what a resilient community should be.

Methodology for The RAFT qualitative assessment

As The RAFT has integrated the qualitative assessment into its broader assessment of community resilience, pairing it with the quantitative scorecard, the process has developed into an effective methodology for engaging diverse perspectives and capturing insights from many sectors and sub-communities within a region. In this section, we provide an overview of how this qualitative assessment is conducted and offer tips for conducting a similar assessment in any community or region.

First, it is essential to recognize that conducting a robust and comprehensive qualitative assessment of resilience will require resources in terms of funding and time for staff to conduct both broad and proactive, targeted outreach to stakeholders, host interviews and focus groups, process data, and synthesize and reflect findings back to the community for their own review and use in resilience planning. It is important not to expect to be able to do this without dedicated funding and at least a few months to execute the assessment, or there is the risk of having narrow input and misrepresenting the broader community's resilience priorities and needs.

It is also important to define resilience for the purposes of your assessment and establish a shared understanding of the term, especially in a focus group setting. For example, The RAFT defines a resilient community as one that is able to anticipate, adapt, endure, and thrive in the face of change, uncertainty, and adversity. When we conducted our qualitative assessments, we shared our definition of resilience with interview and focus group participants before asking them questions about resilience challenges, strengths, and opportunities. As our definition of resilience has evolved, we share that expanded definition with participants in the listening sessions to bring everyone to the common starting point for the assessment.

The recommended steps for conducting a qualitative resilience assessment are:

1) **Define your geographic region and set of communities to bound your assessment appropriately.** Select an area that will be large enough to allow you to work on addressing resilience challenges in the implementation phase. For example, a small town may not be a large enough geographic region if certain operations are managed or decisions are made at the county level. On the other hand, if the region you define for your scope is too broad, it will be difficult to actively recruit and engage stakeholders across the region. It will also require more time and resources to conduct an assessment and advance implementation.
2) **Draft a set of focus group and interview questions that you plan to ask all participants.** While interviews and focus groups can be semi-structured and follow the topics each group brings to the discussion, your analysis needs to include several consistent questions asked of each individual or group. Table 3.1 provides some suggestions for questions you might include in your qualitative assessment.
3) **Begin stakeholder mapping** by creating a spreadsheet or database based on preliminary lists from partners or clients (if available), supplemented with research. If you have partners in a region, they may be able to provide you with a preliminary list of groups they suggest including. However, it is always important to

go beyond the stakeholder list provided by a single entity. Online research and reviewing the member lists of resource councils, watershed collaboratives, and roundtables can broaden your outreach list. Having multiple people on your team contribute to, or at least review, the list will help to mitigate gaps and build a wider, more diverse stakeholder list. Figure 3.4 offers some suggestions for the types of organizations and groups you might include.

4) **Ground truth and expand stakeholder list with key partners embedded locally.** Once you have a draft stakeholder list, share it with your local partners and invite them to suggest additions. For example, if you are working with a local government, have key staff review the list and suggest others you may have missed. If you are collaborating with a regional entity, such as a planning district commission or a nonprofit group that operates across several towns and counties, they may have suggestions for other groups that operate at a similar scale.
5) **Begin outreach to stakeholders, inviting them to participate in listening sessions on resilience in their community.** These can be offered in-person, virtually, or ideally in both formats. If held in person, seek a location that is easily accessible by public transportation, and consider offering meetings both during the day and in the evening to accommodate various work schedules. Ideally, if resources are available, offer an honorarium or other cash or gift card compensation for participants' time. This is particularly helpful and appreciated if you are working with under-resourced communities or asking people to volunteer their time outside a formal job where they are compensated.
6) **Offer interviews to individuals who are unable to participate in focus groups** or who want more privacy in sharing their responses. Some stakeholders may be more candid in an anonymous setting. Offering private interviews can be a way to include individuals who would not feel comfortable participating in a group discussion, or who may not have more than 30 minutes to share their input.
7) **Create and share a survey as a follow-up or alternative option** to individuals or groups who cannot participate in focus groups or interviews. This provides multiple options for capturing insights and acknowledges that different individuals have different capacities and may prefer to engage via different platforms.
8) **Repeatedly solicit suggestions for other individuals and groups to invite to the process** throughout the focus groups

Table 3.1 Sample questions for focus groups and interviews

- A resilient community is one that is able to anticipate, adapt, endure, and thrive in the face of change, uncertainty, and adversity. Given this framework for resilience, how does your organization support resilience in this community? Who do you serve?
- What resilience strengths and assets have you observed in the groups or communities you work in or with?
- Given your organization's work, what people, neighborhoods, and/or populations are identified by your organization as historically excluded and underserved?
- How does your organization serve these populations? What underserved people are not being served by your organization or other groups?
- What strengths, needs, and opportunities in your resilience have been exposed – and for whom? These could range from planning and leadership to infrastructure and health, food, shelter, and wellness. For example, these could be economic or social vulnerability and exposure to climate threats, such as storms, flooding, groundwater intrusion into septic tanks, saltwater intrusion into wells, or an extreme heat event that might impact access to critical services.
- How does your organization work with others to address resilience more comprehensively?
- What other opportunities has your organization identified for increasing the resilience of the people you serve? What ideas have you identified to support increased resilience?
- How might we know this effort to increase resilience is successful, in terms of specific changes that might be visible or measurable? Are there ways of measuring success that would be helpful? In one year? In five years? In 10 years?
- What would your organization need for you to be able to continue to stay engaged in ongoing discussions and networking about community resilience?
- Do you have suggestions for others in the region or your community who should be included in the discussion about community resilience?

and interviews. At the end of each listening session, request that participants suggest others who should be included in the assessment, if you have time and capacity to offer more focus groups and interview sessions. This snowball sampling approach, where existing participants refer or invite additional participants, greatly enhances the breadth of participants and ensures that the assessment reflects the true diversity of the community and groups that may not have an online presence.

9) **Analyze data to identify key themes across interviews and focus groups, and rank the challenges, opportunities, etc. that were raised**

Government, Tribal, and Regional Entities

- Regional planning bodies
- Local government functions (planning, public works, housing, social services, etc.)
- Elected officials
- Tribal governments and representatives
- State agencies (health, emergency management, environmental quality, transportation, housing, etc.)
- Federal agencies (FEMA, NOAA, EPA, HUD, USDA, USACE, NPS, etc.)

Emergency management, Public Safety, and Disaster Response

- Fire and police departments, sheriff's offices
- CERT, Medical Reserve Corps
- American Red Cross and similar nonprofit responders
- Other emergency response or preparedness organizations

Health, Social Services, and Vulnerable Population Support

- Healthcare providers
- Social service organizations
- Community Services Boards, resource councils, aging and disability services
- Shelters, crisis centers, domestic violence services
- Youth and family-serving organizations (YMCA, Boys & Girls Clubs, community centers)

Housing, Transportation, Infrastructure, and Basic Needs Organizations

- Housing authorities, affordable housing groups, homelessness service providers
- Transportation agencies, paratransit and volunteer transportation programs
- Food banks, pantries, farmers markets, community gardens
- Utilities and infrastructure providers

Education, Research, and Knowledge Institutions

- K–12 school systems, preschools
- Colleges, universities, community colleges
- Libraries, museums, nature and learning centers
- Cooperative Extension, research institutes, Sea Grant programs

Environmental, Conservation, and Natural Resource Entities

- Environmental nonprofits, land trusts, watershed associations, riverkeepers
- Soil & water conservation districts, wetlands boards
- Master Naturalist, Master Gardener programs
- Parks, protected lands, and stewardship groups

Economic, Workforce, and Business Interests

- Chambers of Commerce, tourism boards, Main Street organizations
- Industry groups (watermen, farm bureaus, cooperatives)
- Agricultural businesses, marinas, outfitters
- Realtors, hotels, and other business stakeholders
- Large institutional landowners or employers
- Insurance sector representatives

Community, Faith-Based, Civic, and Additional Stakeholders

- Faith communities and interfaith groups
- Civic organizations (Rotary, Elks, United Way, neighborhood associations, volunteer groups)
- Arts, culture, and heritage organizations
- Immigrant and refugee-serving organizations
- Mutual aid networks and informal volunteer groups
- Cultural leaders and tradition bearers
- Climate/environmental justice organizations

Figure 3.4 List of potential stakeholders to consider in conducting a qualitative assessment.

Source: Created by authors.

by the greatest number of participants. Weighting this feedback can be very valuable later in a resilience planning process, by demonstrating which issues are of greatest concern in the community or where there is interest in working collaboratively to solve a challenge.

10) **Be sure to follow up with participants by sharing the findings in a report or presentation.** Be clear about next steps and ways

for participants to engage in the broader community resilience planning process as you move forward with the project.

Lessons learned for practice

1) **A quantitative assessment of a community's resilience needs to be comprehensive, covering multiple dimensions.** It is important to base the assessment on a broad definition of resilience, one that encompasses more than simply emergency response and planning for impacts, to include social and physical infrastructure, natural- and nature-based features, and health and well-being.
2) **Local governments have limited staff, often with frequent turnover, so they face significant resource and capacity constraints**. As a result, they usually do not have the time to complete a resilience assessment or scoring exercise themselves. Clarifying these constraints up front and embedding dedicated external assistance, whether through universities, regional entities, or nonprofit partners, greatly increases the likelihood that assessments will be completed and used effectively.
3) **Any resilience scoring effort needs to be an iterative process** that recognizes that the metrics and questions in the assessment may need to change over time or be adjusted to reflect the specific locality or community you are working with. Periodic pulse checks with locality staff during the assessment phase are important to ensure accuracy and build trust and buy-in. For example, The RAFT scoring team meets with locality planners or other subject matter experts throughout the scoring process to ground-truth their findings. This is critical to earning the community's and local government staff's trust in the overall quantitative assessment.
4) **A full picture of community resilience requires both quantitative and qualitative assessments.** The two processes can be conducted in tandem and be complementary efforts, with the scoring informed by community members' input, and insights gleaned through the scoring providing additional ideas for programs and people to include in the qualitative interviews, surveys, and focus groups. Neither method alone is sufficient for capturing the full complexity of community resilience.
5) **The more inclusive the approach used for a qualitative assessment, the more robust the community resilience**

assessment will be and the broader the knowledge base for the community's selection of resilience actions to pursue. The selection of resilience priorities must be a community-led process, as is the implementation of the chosen actions. A thorough qualitative assessment lays the groundwork for an inclusive, effective process with community members in the driver's seat.

6) **An objective assessment that relies on verifiable processes and evidence** and is conducted by a party that is seen as neutral by participating localities is important for launching effective resilience planning efforts and building community trust.
7) **Effective assessment requires a team with the right mix of skills and knowledge**, including planning, facilitation, legal and regulatory knowledge, data analysis, and community engagement. Without this expertise, scoring may be inaccurate or incomplete. A team approach also builds shared ownership and trust in the results, while improving accuracy and comprehensiveness.
8) **Maintaining and updating assessment tools is time- and resource-intensive.** Adequate funding, staffing, and partnerships are essential to keep tools relevant and ensure that scoring reflects current laws, local conditions, and community needs.
9) **Funding limitations must be recognized early.** If resources are limited, consider narrowing the assessment's scope while maintaining both quantitative and qualitative components, or delay the process until sufficient funding is available to do it more thoroughly. This requires making trade-offs between the need to spark community resilience discussions and increase community awareness of risks and strengths, and the desire to conduct the most comprehensive, inclusive resilience assessment possible.
10) **Creative partnerships, such as with regional planning organizations or local philanthropic organizations, can significantly expand capacity,** especially if you are operating with limited resources. Regional partners often have deep local knowledge and established relationships that can speed material gathering, improve accuracy, and support presentation of results. Partnerships with regional planning organizations have been key to The RAFT's success. The staff of these organizations are familiar with local risks and challenges. They know their member localities and can pave the way for an accurate assessment by introducing a team of resilience assessors to local government staff who can assist with finding materials during scoring and the

vetting of results. They can also provide connections to the local governing boards for presentation of the scorecard results.

Note

1 Rittel, H. W. J., & Webber, M. M. (1973). Dilemmas in a general theory of planning. *Policy Sciences*, *4*(2), 155–169.

4 Small sparks, big waves

Simple projects can have outsized impacts as catalysts

Introduction

The Resilience Adaptation Feasibility Tool (RAFT) process includes a 12-month implementation period focused on completing small projects or initiating longer-term projects to build momentum for continued resilience efforts. The application of The RAFT has shown the importance of prioritizing small projects that can be completed or make significant progress within the year. These early wins not only help move the needle on resilience but also increase awareness of and support for resilience efforts. This chapter offers examples of implementable projects and explains how these small or starter projects set the stage for more sustained efforts to improve equitable resilience outcomes.

Resilience starter projects in education, outreach, research, and training: Examples from the Town of Cape Charles

Opening the door for conversations about resilience that span silos

A charming town with tree-lined streets and front porches, the Town of Cape Charles offers the quintessential small town experience with close to 1,200 residents and the only free public beach on Virginia's Eastern Shore. It is often the first stop for visitors to that area, and it thrives on tourism revenue to supplement its fisheries and seafood industry. Between the expansive Chesapeake Bay, the largest estuary in the United States, and the town's historic homes, homemade ice cream, microbrewery, art shops, and restaurants is a line of protective high dunes.

DOI: 10.4324/9781003596233-4

One challenge identified by town participants at the community workshop was the need to educate town leaders about the topic of resilience and, more specifically, the town's resilience strengths and gaps. When The RAFT 18-month process was completed, the Town Manager shared an unanticipated outcome: The RAFT process had been transformative because the educational presentations about resilience had *normalized* a conversation in the town about resilience and allowed people to openly and candidly share their concerns, knowledge, and hopes. Before this process, attempts to raise issues about resilience reportedly fell flat and went nowhere.

In Cape Charles, the most significant lesson learned was that something as small or seemingly routine as an educational workshop for town leaders could have an impact far beyond what anyone might have imagined. The workshop was an intimate, carefully planned event lasting seven hours, with 15 community leaders taking part. Participants learned about and discussed the concept of resilience, learned from The RAFT Scorecard about the town's resilience strengths and gaps, and then had lively and intense discussions about what they would choose as their resilience priorities for the coming year. In addition, The RAFT team provided an educational presentation about the region's risks to the Town Council, including maps for storm surge and basic inundation projections for 2045 and 2065, to ensure a common understanding of the forecasted future challenges.

As facilitators, yes, we knew the workshop was successful. But we had no idea that the workshop could ever have the impact of creating awareness of the need for resilience planning, as later described by the Town Manager. Now, thanks to the discussions and purposeful gathering of town leaders and experts across silos fostered by The RAFT, we were told the town was finally having open conversations about resilience. Through this, we learned that professional silos are common, no matter how small the community. People all knew each other and wore multiple hats in this small town. Some on the Harbor Review Board also served on the Wetlands Board. The Executive Director of the Historical Society was also the Market Manager for the Farmers Market. The President of the local business association was also the head of the local youth gardening initiative. While they were each connected in different ways through their unique roles, they had never before sat down together to discuss long-term risks and threats to their town's well-being and resilience. This was a big eye-opener for our team: our facilitative and convening role made a difference by creating an intentional space for new conversations about their community's future. Our facilitated conversations also opened a door to ongoing communication and discussions about shared challenges. Resilience had been normalized as a topic to be discussed openly. The Town Council agreed to begin creating a comprehensive Resilience Plan. The Town Manager also shared that the town's preparedness for an imminent hurricane was demonstrably improved because of The RAFT process.

Clarifying policy and authority for management of sand and dunes

In several smaller communities, priorities to support resilience have often required legal and/or policy expertise. As part of The RAFT process, once the community identifies these priorities, The RAFT team identifies who might have the appropriate expertise to address or support the implementation of each priority. The RAFT team is fortunate that one of its university partners offers legal expertise in Virginia's laws and regulations and has been able to provide this support. For some, law and policy issues may not seem like obvious choices for resilience-building priorities. Some might also wonder why a local government would require outside help with law and policy issues; after all, don't they have their own legal counsel? Yet, by listening to the communities and their need to access expert independent opinions on law and policy, we learned that issues that might be easy for well-staffed local governments to accomplish could be significant decision-stopping obstacles for smaller communities that do not have an attorney on staff and do not have the resources to pay private counsel to answer law and policy questions. Some communities we supported had one volunteer Mayor and an all-volunteer Town Council: zero staff, zero legal counsel. Some had only one, two, or three paid staff, each wearing a dozen different hats. Yes, they could ask for support from the surrounding County government or the regional government, but their issues might be put on that legal counsel's back burner. For smaller communities, the ability to gain independent guidance on complex law and policy issues can be transformational, finally enabling sound decision and policy making around longstanding issues.

As anyone familiar with coastal features knows, dunes shift with the winds and weather. They can get smaller or higher, and they can migrate. In Cape Charles, the dunes were becoming a significant worry for the town leaders, as they were shifting, often onto the road. The public beach had been the deposit site for roughly 120,000 cubic yards of dredged material as part of a U.S. Army Corps of Engineers' federal harbor dredging project. This, along with natural movement, had only increased the height of the dunes, creating a surplus of sand that blocked views and created issues on the town's beachside roadway.

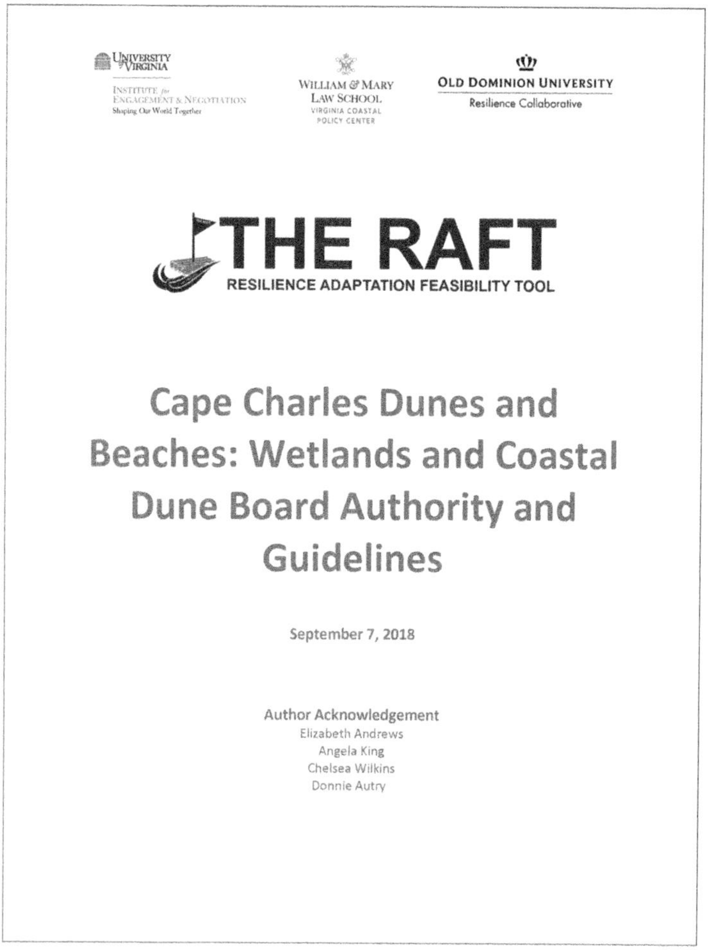
University of Virginia
Institute for Engagement & Negotiation
Shaping Our World Together

William & Mary Law School
Virginia Coastal Policy Center

Old Dominion University
Resilience Collaborative

THE RAFT
RESILIENCE ADAPTATION FEASIBILITY TOOL

Cape Charles Dunes and Beaches: Wetlands and Coastal Dune Board Authority and Guidelines

September 7, 2018

Author Acknowledgement
Elizabeth Andrews
Angela King
Chelsea Wilkins
Donnie Autry

Figure 4.1 Example of policy research projects addressing the dunes problem in Cape Charles.

Source: Created by the authors.

For the Town of Cape Charles, one long-standing question involved the sand and dunes along their coastal road. What could be done, and who had the authority to manage the problem? This had become an increasingly pressing issue with no legal counsel on staff. The law and policy experts on The RAFT team researched and prepared a memo outlining the authority of a local wetlands and dunes board, the authority of the Town, sharing examples from other localities, discussing other relevant state and federal programs, and spelling out options for the Town's consideration. This small research study proved critical, with an outsized impact, as it created a path for the Town Council to have an informed discussion about the range of possible actions. It provided the Town government with the ability to move forward, should it choose to do so.

Outreach and networking for road maintenance

In other situations, policy issues that emerge as important resilience priorities may require more networking than research. Again, the surprise for our team was that these policy issues were relatively small

Figure 4.2 Clogged ditches trap debris and stormwater, causing frequent roadway flooding that could be reduced with regular ditch maintenance.

Source: Photograph provided by Virginia Sea Grant. Used with permission

in scope but similarly so time-consuming that communities with very limited or no staff weren't often able to tackle them without support. An example of this involves road maintenance in light of recurrent flooding, identified by several communities in the Eastern Shore of Virginia, including Cape Charles, as an ongoing and increasingly challenging issue for their resilience. Roads were flooding with increasing frequency, and in turn, this flooding was impacting school bus routes, commuting to work, access to food and medicine, and more.

A number of The RAFT communities felt that their transportation needs were not being prioritized and their questions were not being answered. Could parts of the roads be elevated? Could ditches or swales prevent some of this flooding? Who was the right person to talk to? The state Department of Transportation (DOT)? Or the county Boards of Supervisors who set priorities for projects submitted to DOT? Or the regional government that had provided mapping of flooded roads and other services to the local governments within its region?

When this issue was identified as a common regional resilience priority, The RAFT team agreed to work on it. It took months of work and proactive networking, an email here and there, a call here and there, until finally an understanding of who needed to be at the table came together. We convened an in-person meeting for all interested local governments in the region with the appropriate staff identified by the state DOT – including staff from DOT headquarters, the appropriate regional district office, and the local DOT residency office – and facilitated a conversation about the issues, challenges, options available, and communication channels. Again, this may seem like a simple conversation, yet it required significant time and effort to execute and proved to have an outsized impact, bringing clarity and creating new communication channels between the local governments and DOT.

When there are few or no staff at local governments, the frustrations of reaching out to get answers, trying to get on someone's schedule, and finding the right organization or person who can answer your questions turn challenges into decision-stopping obstacles. Identifying time as one of the most constraining factors and limited resources in communities may seem like a no-brainer. And yet, time and again, it was a surprise to learn that priorities identified by communities often involved seemingly simple tasks that had proved daunting simply because of the time required to execute them and the lack of knowledge of which agency or staff person to contact. A takeaway from these experiences is that practitioners looking to start community resilience efforts should seek partners with familiarity with and contacts in state government or include state agencies in their network of partners.

Through this project, The RAFT was able to grow partnerships between counties and small towns, regional governments, and state agencies. This facilitative role may be a heavy lift, involving numerous calls, emails, and working with community participants to determine when and how presentations are needed, by whom and to whom, to help advance a particular community's goals. Yet it is precisely this intensive investment of time and attention at the ground level that creates opportunities and connections, builds trust and networks among community members, and enables them to continue working together. A senior planner for a regional government and other Implementation Team members repeatedly emphasized how much The RAFT helped strengthen their relationships with one another and jump-start community-wide discussions on resilience and specific projects that could be funded.

Lessons learned for practice

1) Never underestimate the power of a small, focused educational presentation or workshop to transform conversations.
2) Even the smallest of communities have silos of expertise and communication. People wear many hats in small towns, and because of this, they are stretched to the limit. So, any expectation that people will communicate with each other or know about each other's work, simply because it is a small town or rural county, is not necessarily true.
3) In under-resourced communities, your ability to coordinate and facilitate one or more gatherings at no cost to the local government or community will likely make the difference between an event happening there or not.
4) Most people are grateful beyond measure for coordination and facilitation support, and assistance in bringing important information to the community and opportunities to network across silos to support their community.
5) Under-resourced communities may not have easy or affordable access to policy or legal advice on what they can and cannot do related to improving resilience. You can seek assistance from your

local college or university. Faculty and students in political science, policy, or law programs may be able to help with policy or legal research, and this initial collaboration could lead to a long-standing partnership that you can facilitate between the community and faculty members.

6) Legal research and guidance, even on what may seem like a relatively simple set of questions, can make the difference between a problem getting resolved and addressed or being put on hold for years.
7) Many resilience challenges go unaddressed because there is a lack of clarity around what is or is not possible. Holding space for discussion and facilitating access to experts create clarity about possible pathways of action, leading to outsized value in helping your community be effective in addressing its resilience challenges.
8) Communication is critical to ensuring a common understanding of the resilience issues at hand. Enabling and facilitating representatives from various groups, such as local, regional, and state governments, to meet in a room and speak with one another about ongoing, persistent issues can be transformative. For example, smaller local governments with few staff may not be aware of existing regional government studies and relevant committees that could help them.
9) State agencies operate in large, often complicated sets of rules and regulations. Rather than immediately attempting to revise a statewide program or approach, it is important first to create a forum to build understanding, open channels of communication, provide local governments with the information they need about avenues of support and funding, and assist the locality in considering which avenues to pursue.

Resilience starter projects in planning and mapping

Shoreline erosion and stormwater planning: Examples from the Town of Colonial Beach

West of the Eastern Shore, across the vast expanse of the Chesapeake Bay, the rural region of the Northern Neck offers more examples of how small projects can provide outsized impacts. Specifically, small planning and mapping projects can be transformative in advancing a community's resilience, and often these small projects might be undertaken by university faculty with their classes, at low or no cost to the community.

Early in The RAFT implementation process with the Town of Colonial Beach, a small waterfront town, we began meeting with the Deputy Town Manager and the new GIS staff technician to refine further the Town's resilience priorities and specific mapping needs. Based on these discussions, the Town decided to focus on two priorities: stormwater issues and shoreline erosion, including exploring opportunities for living shoreline remediation along its waterfront.

To assist the Town of Colonial Beach in meeting these resilience needs, we coordinated support from several partners to provide pro bono expertise and analysis. We first recruited a university planning professor who was also the founder of a nonprofit, the Green Infrastructure Center. The professor offered to have her class assess green infrastructure opportunities for addressing an array of resilience issues in Colonial Beach. The following semester, her class worked on design options for specific sites to become functioning green spaces for community gatherings while also addressing flooding concerns. The students also conducted a review of the Town's codes and ordinances using the Green Infrastructure Center's "Trees and Stormwater Audit Tool" to identify places where green infrastructure could be more intentionally prioritized to mitigate stormwater impacts. Students presented their assessment results, site plans, and recommendations to locality staff at the end of the academic year. Another planning professor conducted an assessment of climate adaptation opportunities for the Town. This assessment identified specific ways to address the Town's resilience challenges and includes specific references to shoreline erosion and stormwater.

Each of these examples demonstrates the rich potential of university faculty who bring expertise across different aspects of community resilience. In each of these instances, the faculty members used their class for the dual benefit of giving students a meaningful, real-world community project to work on and learn from, while also providing a customized, valuable service at no cost to the community. And, above all, each of these projects addressed real issues and delivered immediate, outsized impact – a new park design, a green infrastructure framework, and new language for the Town's comprehensive plan that could ensure long-term integration of resilience planning into the Town's future.

A third professor mapped the Town's shoreline erosion issues. He and his students conducted a drone flight over a part of the Town shoreline that staff and residents had prioritized as the area of greatest concern. This data, in combination with bathymetric data collected

Figure 4.3 Shoreline data were collected to map shoreline erosion using drone and hydrone.

Source: Photograph by Tom Allen. Used with permission.

by hydrone, was used to develop a "State of the Beach Report" for the Town.

Several years later, the Town of Colonial Beach successfully applied for grant funding of over $100,000 to develop a formal Resilience Plan. In the grant proposal and cover letter, The RAFT process and resources and relationships that it facilitated were cited as critical to building the networks, partnerships, and understanding of resilience needs.

Mapping community green infrastructure: Example from the City of Petersburg

Below the Virginia state capitol, right off the major east coast north-south Interstate I-95, lies a small city with rich history, Petersburg, that today is seeking to recover from a period of decline and to grow a more resilient economic base. With a majority African American population, the City became an early transportation hub in the days of the canal system and later with the advent of railroads, serving as a major connecting point for goods east and west. To this day, it remains a major transportation hub for rail and the Interstate highway system. The rich history of Petersburg runs the gamut, from being home to the largest free black settlements in Virginia, to being the site of the longest siege in American military history, ultimately leading to the end of the Civil War with Robert E. Lee's surrender in 1865. The City seeks to integrate its many histories with a bold vision for a thriving future and is seeking support from the State and elsewhere to realize that vision.

During The RAFT process, the community identified two significant priorities that would contribute to Petersburg's vision for a more resilient future: landscape plans and standards to protect and expand its green spaces, and a plan to ensure that residents have access to goods and services via their trolley system.

As the community began brainstorming ways to implement these goals, it became clear that the City would benefit from a green infrastructure plan that maps existing green spaces that should be protected and identifies spaces that could become green. The RAFT team enlisted the same university faculty member with the Green Infrastructure Center, whose class had supported the Town of Colonial Beach in its resilience work. Working with the City and local nonprofit partners, the faculty member and her students created a green infrastructure plan that centered Petersburg's priorities: parks and recreation, trees, water, culture, and history. A different set of students then developed conceptual designs for four specific sites in the City that could be transformed into green spaces.

We worked closely with city staff throughout this collaboration to ensure that staff and contractors could integrate these mapping and design products into the City's comprehensive plan. What's more, the Green Infrastructure Center worked with city staff to use these green plans and conceptual site designs to secure a state grant to conduct a tree canopy assessment and draft a Tree Ordinance to advance urban tree canopy coverage.

In turn, this led to an in-person community educational workshop on street trees and urban greening. Participants learned about the connections between green infrastructure and health, as well as the City's goals for tree canopy and green spaces. Community feedback at this event helped to shape the City of Petersburg Strategic Tree Canopy Plan.

Never had we anticipated or imagined this cascade of events and accomplishments. It would not have been possible without matchmaking Petersburg with an interested university faculty member, her classes, and her small nonprofit center. Beginning with a simple goal of creating a green infrastructure plan, so much more was accomplished, including designs for four sites, funding for a tree canopy assessment and a tree ordinance, and a community workshop to engage residents in learning how to become part of protecting and growing their green spaces.

Mapping mobility to increase accessibility through the community trolley: Example from the City of Petersburg

A second resilience goal of Petersburg evolved beyond its original vision for increasing accessibility through the community trolley, again thanks to match-making the accessibility goal with a university professor of urban planning. Working with the community, the professor's students used mapping tools to research the community's needs for micro-mobility (the first and last mile), mobility, and access to food, green spaces, and broadband. To do this, they identified activity centers and likely destinations, including businesses, schools, the university campus, historic military Fort Gregg-Adams, and other

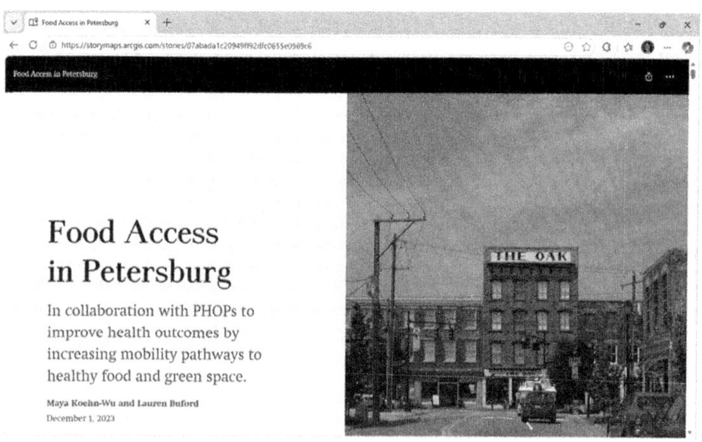

Figure 4.4 Screenshot of the Food Access in Petersburg StoryMap.

Source: Created by authors from the StoryMap by Maya Koehn-Wu and Lauren Buford. Used with permission.

locations. They then developed recommendations for safe and accessible e-scooter routes.

In addition, the students supported an ongoing walking and road safety audit by highlighting, using an ArcGIS StoryMap, resident challenges in accessing healthy foods and green spaces. This StoryMap aims to provide the City with a visual tool to prioritize new transit routes and site new parks, community gardens, food banks, and farmers markets, thereby maximizing access for under-resourced residents.

Next, building on the student work, The RAFT team developed and conducted a community survey for the City to learn about the best routes for the free trolley to provide residents with easy access to green spaces and healthy foods. This survey was needed for the City to secure funding for expanding its trolley service.

Last, The RAFT team played matchmaker again, this time connecting an engineering professor eager to give his students a real-life project in which they could apply their GIS skills to real-world transportation challenges. The GIS students analyzed the community survey results to develop a proposal for trolley routes that would best address resident needs, particularly in low-income neighborhoods with limited transportation options. They also provided cost estimates for these trolley routes. This work enabled the City to successfully apply for operational funding to launch and pilot the free trolley along the proposed routes.

From two small project beginnings came unimaginable gains – funding for the Tree Canopy Assessment and Plan, a model ordinance, a Green Infrastructure Framework, conceptual green designs for four specific sites, a StoryMap of places and activities that residents want to access, a plan for a free trolley that included routes and cost, and funding to launch and pilot the free trolley. Without matchmaking projects with university professors who had relevant expertise and were eager to incorporate real-world projects into their classes, none of this might have been possible (see Chapter 7 for a discussion of partnering with universities and other entities). And yet, for the effort it took to recruit these professors, the outcomes were many times greater.

Lessons learned for practice

1) Sometimes a small step is more important than a big step, as it can offer an easier win and catalyze community enthusiasm and local government support for continuing to pursue resilience projects.

2) Smaller planning and mapping projects (such as projects that can be accomplished by a university class in one semester) can offer outsized impacts, as they can provide a community with the map, or design, or planning language, or other framework that enables it to move forward, apply for funding, seek approval, enlist support, or take the next step in improving their resilience. The effort it takes to match a project with a university professor with relevant expertise can pay off in hefty dividends.
3) Small projects also can have a cascade effect, with one accomplishment leading sequentially to a series of unexpected additional projects and outcomes. Matching a community's needs with relevant faculty expertise can go a long way toward achieving outsized impact and allowing that faculty member or their institution to continue providing support or pursuing new projects directly with the community.
4) Faculty at universities are a largely untapped powerful resource for supporting community resilience. They may not know how to connect with or create relationships with communities, but they may be excited to use their class to undertake a specific project for a community. In light of this enormous, untapped resource, we are expanding our networking and matchmaking capacity by building a statewide roster of community-engaged faculty.

Simple solutions to support residents

Refrigerator magnets with critical emergency information: Example from the Eastern Shore of Virginia

On the Eastern Shore of Virginia, the population swells with visitors as the weather warms, many of whom stay in rental properties through Airbnb, Vrbo, or other independent property owners. During this annual cycle of high visitor population is precisely when hurricanes and nor'easters are most likely to create dangerous flooding situations and the need for evacuation.

During community discussions about ways to increase resilience, we heard repeated concerns about the need to educate and inform people – and most especially those who are not year-round residents – about what to do and where to go in the event of an emergency. Flyers in mailboxes had been tried, but at rental units they sometimes weren't picked up or were blown away by storms, so there was no guarantee the information was reaching occupants. Out of this came an unusually simple idea for a refrigerator magnet that would remain posted inside, out of the weather, showing the evacuation route and providing numbers and locations for medical or other kinds of emergency needs. Everyone loved this elegant solution! All residents and independent rental properties would have refrigerators, and refrigerator magnets would be literally in everyone's face, readily available in an emergency.

The RAFT team worked with community participants to design the refrigerator magnet, ensuring it included all the desired information. One of The RAFT university partners was able to provide the small amount of funding needed to print the magnets. Although the magnets were originally created for one locality, when other local governments learned of them, each requested a version for its residential homes as well. In the end, thousands of magnets were created and distributed to both counties and all four towns. This remains one of The RAFT's easiest and highest impact solutions for increasing resident emergency preparedness, a vital element in building community resilience.

Short-wave radios for communicating during emergencies: Example from the Town of White Stone

Another example of an unexpectedly simple solution for residents comes from the tiny Town of White Stone, nestled in the Northern Neck Peninsula. Discussions with this town revealed a different kind of challenge faced by its large contingent of older, retired residents. With poor internet and cellphone connectivity and many residents who were not raised in the digital age and had limited proficiency with computers, communicating with and among residents was a key challenge. During discussions with The RAFT, the town manager surfaced the idea of short-wave radios.

The idea was immediately embraced as a solution that worked for this particular town, given its specific demographic and communication challenges. Emergency managers could easily communicate with residents via radios in the event of an emergency, and the radios were simple enough for all residents to operate. We approached a local community foundation for assistance with this project, and the foundation provided funding to purchase 500 radios, which were distributed to elderly residents.

Lessons learned for practice

- Never underestimate the power of a simple solution. Sometimes the simplest solutions may be the most effective and the least expensive.

- Small is beautiful. Simple is beautiful. A long, complex communication plan was not needed for either the Eastern Shore or the Town of White Stone, at least in this particular case. If they had been sold on the need for such a plan, their challenges still would not have been addressed.
- Cookie-cutter approaches are not helpful, although models from other places can serve as inspiration or a starting place for discussion. It is important to listen to each community's needs and customize the solution for that community. The Eastern Shore community would not have solved its problem with radios, and the Town of White Stone would not have solved its problem with magnets.

Simple solutions to support businesses and provide technical assistance

Small Business Resilience Self-assessment and Guide

Not far from the Rappahannock River and in the heart of the Northern Neck, the small Town of Warsaw, with fewer than 2,000 year-round residents, is on a path to revitalize its Main Street and downtown districts. Participants in the Town's community workshop identified business resilience as a concern, and it was prioritized as an implementation project on the Town's Resilience Action Checklist.

To address this need, we developed the Coastal Virginia Small Business Resilience Self-Assessment and Guide, designed to help small businesses evaluate their resilience to a wide range of risks and develop a practical plan of action to strengthen their ability to withstand and recover from disruptions. The development of the self-assessment and guide was informed by direct engagement of small businesses in the Town of Warsaw, who participated in interviews and focus groups, and provided feedback via email. These business owners consistently expressed the need for tools that were accessible, actionable, and tailored to their conditions.

They identified challenges such as difficulty assessing hazard exposure, limited time or expertise to plan for emergencies, uncertainty about insurance needs, difficulty hiring qualified employees and transitioning business ownership, and the operational strain of adapting to broader changes such as e-commerce and digital marketing. These insights helped shape a self-assessment and guide that addresses real-world obstacles faced by small businesses in coastal Virginia, such as continuity planning, staffing shortages, cybersecurity vulnerabilities, and succession needs.

The self-assessment organizes business resilience into four major areas:

1) **Business planning** – Focuses on foundational planning activities such as developing a written business plan, preparing for digital marketing and online sales, and establishing succession and exit strategies. These elements help ensure long-term stability by strengthening strategic direction and preparing the business for transitions and market changes.
2) **Emergency and disaster planning** – Guides businesses through risk and vulnerability assessments, preparedness planning, continuity of operations, and communication strategies. This equips businesses to anticipate hazards, minimize disruptions, and recover more quickly from emergencies such as flooding, hurricanes, or power outages.
3) **Human resources and staffing** – Addresses hiring, training, onboarding, and employee retention to ensure that businesses have the workforce capacity needed to operate effectively, especially during or after disruptions. Strong human resources practices help maintain business continuity and support organizational resilience in the face of unforeseen challenges.
4) **Insurance and protection** – Covers understanding insurance needs, maintaining up-to-date coverage, protecting physical assets, and strengthening cybersecurity. This helps businesses manage financial risk and safeguard critical resources before, during, and after disruptive events.

Alongside the self-assessment that helps small business owners identify challenges, a structured set of resources, recommendations, and worksheets guide business owners through the process of building long-term sustainability. Additional worksheets addressed key topics that business owners in Warsaw determined to be essential topics: digital marketing (highly relevant given the COVID-19 pandemic environment in which The RAFT implementation phase occurred), exit planning, and succession planning.

The self-assessment and guide was made available online for small business owners and managers to download and use. While the self-assessment and guide were developed with input from business owners in Warsaw, the content and resources are broadly applicable to small businesses across coastal Virginia, and the resource was widely distributed to communities beyond the Northern Neck.

Small business resilience workshop

Once the Coastal Virginia Small Business Resilience Self-assessment and Guide was developed and made available in electronic format, we also held workshops for small businesses to help them conduct their self-assessment and develop next steps to improve their resilience and business continuity. We worked with regional planning agencies to offer the workshops, which were supported by GO Virginia, a state-funded initiative administered by the Virginia Department of Housing and Community Development that strengthens and diversifies Virginia's economy.

The Small Business Resilience Training Workshop was designed to introduce participating business owners and managers to the growing risks facing small coastal businesses and the importance of building resilience in the face of disruptions such as flooding, power outages, extreme weather, cybersecurity threats, and supply chain challenges. Participants completed the resilience self-assessment and discussed their businesses' vulnerabilities. They then worked through self-assessment results to develop a personalized action plan. The workshop provided resources on continuity of operations, emergency communication, risk and vulnerability assessment, succession planning, and digital marketing and e-commerce, all topics identified as priorities by small businesses in the region. By the end of the workshop, each business left with a clearer understanding of its resilience gaps, a prioritized action plan, and, hopefully, confidence in the next steps needed to strengthen preparedness. The workshop reinforced that resilient businesses help create resilient communities, making these skills essential not only for survival but for supporting the broader regional economy.

Designing a resilient harbor: Example from the Town of Onancock

When we worked with seven localities on Virginia's Eastern Shore, the Town of Onancock included an item on its Resilience Action Checklist to "Design a resilient wharf waterfront." In particular, town staff noted that their wharf needed repairs to the bulkhead and boat

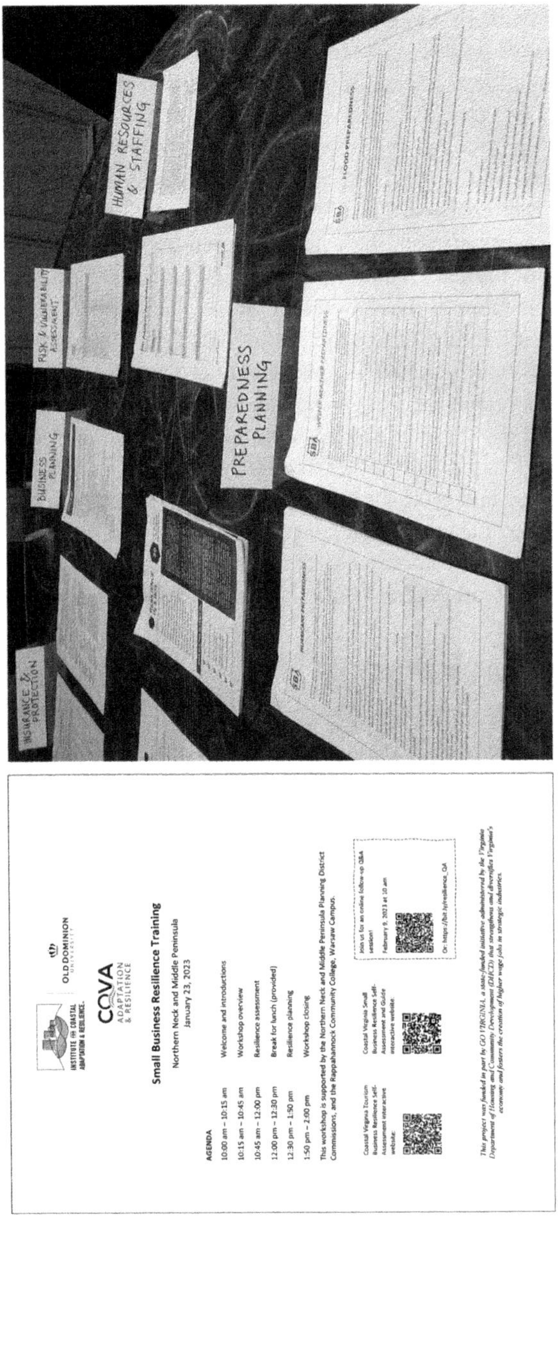

Figure 4.5 Small Business Resilience Training Workshop agenda and resources.

Source: Flyer created by the authors and photograph by Juita-Elena (Wie) Yusuf.

ramp and emphasized that drainage could be improved to prevent flooding of the harbor master's parking lot. The harbor was a particularly important cultural and commercial asset to the community, and protecting it as an asset was of primary concern to residents.

During the year of implementation, The RAFT team facilitated a relationship between the Town of Onancock and a Virginia Sea Grant Fellow who worked with an architecture and engineering team to address this concern. The Virginia Sea Grant Fellow visited the harbor and, working with the team, designed flood control measures for the harbor master's parking lot, including a duckbill valve to prevent the drain from backwashing in high tide. This was at no cost for the community. By the end of the year of The RAFT implementation, the Town of Onancock had a shovel-ready project and, when Port of Virginia grant funds became available, successfully applied for and was awarded the funding needed to build the project and install the duckbill valve to prevent further flooding.

At the year-end workshop for all localities participating in The RAFT on the Eastern Shore, representatives from the Town of Onancock shared the importance of this small but vital fix, describing it as "the most productive, least intrusive and most cost-effective

Figure 4.6 Flooding at the Onancock wharf.

Source: Photograph provided by Virginia Sea Grant. Used with permission.

means in which to protect our town's harbor." A follow-up interview was conducted nearly four years after The RAFT application on Virginia's Eastern Shore, and in this interview, a town Councilwoman who had participated in The RAFT highlighted that she liked participating in The RAFT process because it moved them beyond debating climate impacts on the Town and instead forced them to think about how to deal with the specific impacts.

Resilience mini and micro grants

One of the benefits of working at the regional level is the potential to secure funding from regional place-based foundations interested in advancing resilience across the region. When The RAFT worked with two adjacent regions comprising ten counties in both the Northern Neck and the Middle Peninsula, we had the unique opportunity to partner with a place-based family foundation that welcomed the chance to support a mini and micro grant program to implement the region's resilience priorities. Provided with a pot of $150,000, we offered local governments, tribes, regional organizations, and community and nonprofit organizations the opportunity to apply for small grants between $5,000 and $25,000. Submission requirements were intentionally minimized to eliminate barriers for organizations with limited resources and few paid staff. We also minimized reporting requirements, enabling awardees to provide reports verbally and with minimal written language, thereby reducing their administrative burden.

This was a first for The RAFT. We usually apply for grants, not distribute them, so it required some new thinking on our part. We didn't know what projects might be proposed or which of the community's identified resilience priorities might be addressed. We experienced a number of surprises, perhaps the most important of which was that, when given an opportunity, communities will think very holistically and broadly about resilience. Yes, they considered flooding and infrastructure, but their grant proposals also reflected a broader understanding of resilience to include food, medicine, historic preservation, economic opportunity, and culture. Overall, 14 grants were awarded to advance resilience in the two regions. Each project included a strong community involvement component, was cost-effective, and contributed to social and economic success.

Based on the strong success of this pilot grant program, the regional place-based foundation decided it would be important and effective to continue this method of supporting resilience-building. It worked to create a special fund, managed by a local community foundation, dedicated to supporting community-driven resilience projects. As of this writing, a representative of The RAFT team continues to serve on this fund's grant review and selection committee.

We view this pilot micro and mini grant program as one of our greatest successes because it enhanced The RAFT's traditional process with material funds to actualize and implement community resilience priorities. The flexibility in funding and short turnaround between applications and awards, made on a rolling basis, allowed for a diversity of creative project ideas and responsiveness to community needs. Implementing a micro and mini grant program like this in any region or set of communities requires funders willing to take innovative approaches, invest in flexible funding structures, and channel funds to small community organizations with limited resources. We encourage planners, as well as large community organizations and NGOs, to consider approaching their local community and place-based family foundations with this model to catalyze ground-up engagement in building resilience and achieving outsized impact.

Broadening resilience to include health, housing, and other aspects of well-being

When The RAFT first began, a common question from local government was: "What does health and well-being have to do with resilience? We have to deal with flooding and the changing shoreline." Another question was: "How can we deal with these environmental issues when the most pressing challenge is our economy? We don't even have enough money to fix our school to have an emergency shelter for our community!" These questions arose because The RAFT Scorecard reflects a deliberate decision to approach resilience as a three-legged stool that, like sustainability, is effective only when all three environmental, economic, and social legs are present.

Nearly five years after we launched The RAFT, community participants were not only beginning to embrace a broader view of resilience but were also prioritizing related actions. Not far from the state capital of Richmond, in the rural county of King and Queen, known for its forestry, many waterways, and fewer than 10,000 residents, community participants prioritized resilience actions that fell outside of those typically identified. During The RAFT implementation process, community leaders and nonprofit participants talked about their elderly residents who were shut-in and wanted to find ways to deliver food and medicine *to* them, not just transport them to places where they could get their food and medicine. They also talked about food security and giving people a way to become more self-sufficient by creating the first community garden to enable people to grow food for their own table. These became key priorities on the Resilience Action Checklist.

Both of these priorities were ambitious and complex. We were initially concerned that creating a network of community gardens would be difficult, yet the presence of one strong, passionate advocate led others to support the idea. As a result, a new nonprofit organization, Just Harvest, was established and new collaborations between this group and existing institutions working on food security regionally were strengthened The RAFT team's role, in this case, was facilitative – identifying and connecting relevant experts with the community leaders who wanted to create new community gardens and link existing ones in a broader system of food distribution to populations in need. Slowly but surely, this approach worked, and new partnerships and community gardens began to take shape.

Two mini grant applications supported the establishment of new gardens that were part of this integrated regional food security strategy. The first was a garden that included two in-ground beds and one grow tunnel to grow organic produce, including butternut squash, corn, cucumbers, tomatoes, yellow onions, zucchini, kale, okra, bell peppers, and many more vegetables. The garden distributed approximately 1,500 pounds in its second growing season through different local organizations.

A second mini grant funded Just Harvest's creation of a community garden in neighboring Essex County, where harvest was distributed door to door to 15 households. This project also expanded into another adjacent county, creating an additional garden with an outdoor classroom to provide training and make connections between food, environment, culture, and history.

Another mini grant application came from the Mattaponi Indian Tribe and Reservation, seeking funds to complete repairs to their fish hatchery, which had long been part of their culture. Once repaired with new decking, a well pump, and HVAC, the hatchery could open for the spring American Shad season. The hatchery would contribute to the Tribe's economic sustenance in addition to the Tribal community's food supply.

This repair proved to be more complicated and challenging than anticipated. Though it took longer than the projected timeframe, ultimately it was successfully completed. Moreover, the grant helped the Tribe establish a brick-and-mortar Environmental Office staffed by two full-time employees to continue environmental resilience and capacity building projects that would have a lasting impact on the Tribal community.

Housing was another resilience priority that began to emerge from the mini and micro grant program. One regional nonprofit organization was awarded a mini grant to financially assist 23 households (49 individuals) in attaining permanent housing. Another mini grant was awarded for an emergency shelter group to expand its Motel Voucher Program for residents experiencing homelessness. Prior to the mini grant, people would have had to travel to a neighboring county to access a shelter. The mini grant supported 131 weekly motel vouchers for 13 households experiencing homelessness over a six-month period. Even though some state and regional agencies may not see housing as a fundable resilience priority, we continue to see it as a resilience priority in the communities where The RAFT works.

Examples of other types of non-traditional resilience-building activities that were supported through micro and mini grants include:

- Supporting the preservation of a historic Rosenwald School to serve as a community center. The grant provided for exterior doors custom-manufactured to be consistent with the historic doors used in the original school, to provide a good first impression of the building for visitors, and also to ensure building security and safety.
- Creating a social media site by a local Fire Department to foster monthly business networking to build partnerships, hosting resilience training sessions for fire departments across the region, and conducting a community assessment to support a grant that would pay for the training of first responders.
- Teaching fire safety at a local county fair, events with the YMCA, County Schools, and the annual trunk or treat event in a local town.
- Purchasing technology for an Emergency Operations Center and engaging stakeholders to develop a one-year plan to rewrite the County's Emergency Operations Plan.
- Creating a community-based social marketing pilot project for shoreline management.

- Creating an informational campaign to educate residents and homeowners about failed and failing septic systems, and how and why they need to be fixed.
- Funding septic pump-outs that otherwise would not occur, with a total of 65 septic systems successfully supported across six counties.

We encourage planners and community-based organizations to consider these non-traditional resilience priorities as equally valid and important elements of building community resilience. These types of actions can excite and catalyze organizations to begin seeing themselves as having agency in their community's resilience and viewing resilience as something they can help grow.

Lessons learned for practice

1) Small micro and mini grants of $5,000 to $25,000 can be powerful motivators for local organizations, and may be all that's needed for them to accomplish important resilience goals. Community groups seeking to undertake resilience building should explore opportunities to partner with local community foundations to fund a pilot grant program.
2) Small mini grants do not have to be complicated and, in fact, should be made as accessible and easy to manage as possible. Reduce project reporting to paragraphs and/or verbal conversations to lower the administrative burden associated with grant funding.
3) When allowed and encouraged, communities will naturally begin to think more holistically about their resilience, considering access to food, water, medicine, and preservation of cultural heritage. While flooding, infrastructure, and associated policies are all vitally important for physical resilience, so are other aspects of community for weaving a fabric of emotional and spiritual connectedness that sustains people through crises.

5 Weaving resilience into the mainstream

Integrating resilience and equity across different facets of community planning

Introduction

Resilience spans the whole community. Resilience planning is most effective when embedded in the everyday and regular processes of local policy decision-making and planning, rather than treated solely as a separate initiative. A decade of applying The Resilience Adaptation Feasibility Tool (RAFT) to support resilience in Virginia communities has shown that formal plans incorporating resilience and equity can provide explicit guidance, prioritize actions, and help communities secure funding. This chapter explores how resilience can be mainstreamed into existing planning instruments, such as comprehensive plans, land use plans, capital improvement plans, hazard mitigation plans, and regional healthcare plans, and offers practical guidance for building resilience plans that leverage these various efforts. Drawing on lessons learned from The RAFT's application in different communities, this chapter demonstrates how communities can integrate resilience into planning processes without reinventing the wheel, while ensuring equity remains central to decision-making.

What does mainstreaming resilience mean?

Mainstreaming resilience means integrating resilience concepts into existing plans, policies, and projects[12] rather than simply creating standalone resilience plans. This approach avoids duplication and policy silos by embedding resilience into plans that localities and communities already update regularly, such as comprehensive plans, hazard mitigation plans, and neighborhood preparedness efforts. It also addresses interdependencies between hazards and acute

DOI: 10.4324/9781003596233-5

shocks (such as hurricanes or infrastructure failures) and chronic stressors (such as poverty, aging infrastructure, and racial inequities). Mainstreaming resilience centers equity by incorporating justice into planning processes and ensuring inclusion of historically excluded and underserved community members. For many under-resourced communities, framing coastal hazards within the broader context of social and economic resilience – for example, jobs, healthcare, and food security – is essential. A systems approach that spreads resilience across these aspects of the community is critical because environmental, economic, and social dimensions collectively shape a community's ability to thrive.

Mainstreaming resilience allows broadening the conversation beyond physical hazards to include social and economic well-being. For example, improving food security through comprehensive planning that addresses food deserts or addressing mobility and access issues in transportation plans becomes integral to coastal resilience planning. This approach acknowledges that resilience is multidimensional, encompassing environmental, infrastructural, social, and economic needs. It also emphasizes interdependencies between shocks and stressors, such as recurrent flooding and persistent structural inequalities. By embedding resilience throughout various planning instruments, such as comprehensive plans, hazard mitigation plans, zoning ordinances, capital improvement plans, and transportation strategies, communities can ensure consistency and coordination across sectors and functions.

Mainstreaming also addresses capacity challenges faced by localities with limited staff and technical capacity. For many under-resourced communities, economic and social resilience are often more urgent priorities than coastal hazards alone, and mainstreaming helps connect agencies such as social services and health departments into resilience networks to ensure that vulnerable populations receive targeted support. Mainstreaming resilience ensures that it is not an afterthought or isolated initiative, but a guiding principle embedded across the fabric of community planning and governance.

The RAFT as a resilience connector

The RAFT approach provides a process and structure that serve as the connecting tissue across separate plans, consistently incorporating resilience and equity. The RAFT three-step process begins with a comprehensive assessment of resilience using quantitative and qualitative methods. The RAFT Scorecard evaluates five dimensions through which resilience can be addressed: policy and leadership, risk assessment and emergency management, infrastructure resilience, planning frameworks, and community engagement and well-being. The qualitative approach explicitly addresses equity and how resilience and equity connect at the community's grassroots level. Both assessments highlight gaps and opportunities across policies, planning, infrastructure, and economic and social systems. For example, The RAFT Scorecard might identify that zoning ordinances and the comprehensive plan lack integration of updated flood mapping and risk analysis. Alternatively, the qualitative analysis may surface disconnects between how community members seek information and how the locality provides information.

Beyond offering an assessment of where and how resilience is missing from current policy and planning instruments, The RAFT further builds on existing plans by creating an Opportunity List that draws from the scorecard results, comprehensive plans, hazard mitigation plans, and capital improvement plans, combined with grassroots input from the qualitative assessment. This list respects the work already done by localities while incorporating community priorities identified through the qualitative assessment's focus groups and interviews. For example, locality staff often ground-truth priorities from the qualitative assessment, such as broadband access in rural counties, which may not appear in hazard mitigation plans but are critical for resilience. The Opportunity List provides a starting point for communities to identify priority projects for their Resilience Action Checklist. During regional community workshops, The RAFT team facilitates consensus-building through the Resilience Action Checklist, which includes locality staff-selected priorities and community-driven actions. This iterative process ensures buy-in from local government while elevating grassroots concerns, creating a foundation for resilience plans that can support future funding applications.

The RAFT's approach demonstrates that resilience planning does not require starting from scratch. Instead, it leverages existing opportunities and requirements to integrate resilience and equity into ongoing processes. In Virginia, the localities we work with are required by state law to evaluate their comprehensive plans every five years. They are also expected to update their hazard mitigation plans every five years to ensure eligibility for federal disaster funding.

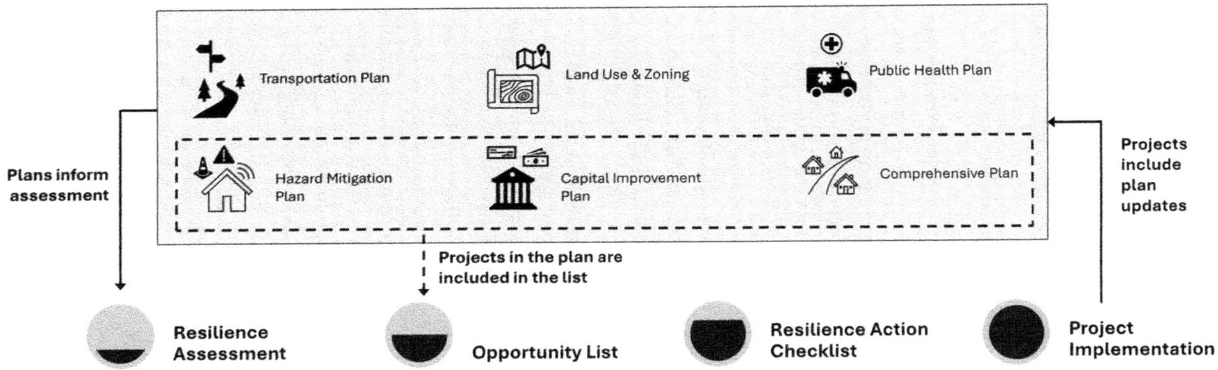

Figure 5.1 The RAFT process and mainstreaming resilience.
Source: Created by the authors.

Similarly, the State has required that localities have resilience plans to be eligible to apply for selected state funding through the Community Flood Preparedness Fund. By connecting resilience and equity as a continuous thread across these efforts, The RAFT helps communities overcome policy silos and build capacity for holistic resilience.

Equity and participation in the planning processes

When people see their voices reflected in plans, they perceive the process as fair and inclusive. This increases buy-in and reduces resistance to implementation. Equity can be incorporated into planning processes by ensuring that all community members, especially those who are historically marginalized or disproportionately impacted by hazards and stressors, have a voice in shaping policies and decisions.

Equity in planning goes beyond equal treatment; it requires addressing systemic disparities and ensuring that resilience strategies do not exacerbate existing vulnerabilities. For example, comprehensive plans can integrate equity by including social vulnerability mapping alongside flood risk assessments, identifying neighborhoods with limited access to transportation, healthcare, or economic opportunities, and including residents from these neighborhoods in the planning process. This allows planners to prioritize investments in areas where residents face compounded risks from hazards and socioeconomic challenges, and to do so with the knowledge of and input from those residents.

Hazard mitigation planning offers another opportunity to embed equity. Traditionally focused on physical infrastructure, modern hazard mitigation plans can incorporate strategies that address social and economic resilience. The RAFT community workshops and Implementation Teams often include social service agencies and grassroots organizations to ensure that community priorities, such as improving evacuation routes for low-income neighborhoods or ensuring continuity of healthcare services, are reflected in resilience actions. These approaches operationalize equity by combining technical data with community-driven insights, ensuring that resilience planning benefits all residents, not just those with resources or political influence.

Effective equity integration involves participatory processes that remove barriers to engagement. Holding meetings and events at times and locations convenient to specific groups of residents enables targeted engagement. Providing childcare, selecting meeting sites with public transportation, and using accessible meeting formats enable underrepresented groups to participate meaningfully. By embedding equity into planning processes through the use of local knowledge and data and inclusive engagement, localities can create resilience plans that are both just and sustainable.

Box 5.1 Governance explained

Governance is the way decisions are made and carried out in an organization, community, or society. It includes the rules, processes, and relationships that guide how power is used, responsibilities are shared, and accountability is ensured.[3]

Participation is not just a procedural step in planning. It can be a transformative approach that embeds equity and resilience into the very fabric of governance. Traditional planning processes often rely heavily on technical experts and formal decision-makers,

Table 5.1 Tips and recommendations for embedding equity in the planning process

- **Engage trusted messengers to reach vulnerable populations.** Community leaders, faith-based organizations, and local nonprofits can help increase participation and ensure equity in resilience planning. These trusted messengers help overcome distrust and language barriers, making engagement more inclusive.
- **Combine technical assessments with grassroots input.** Use metrics such as from The RAFT Scorecard, the PREP Tool, CDC's Social Vulnerability Index (SVI), Census Bureau's Community Resilience Estimates (CRE), and FEMA's Community Resilience Challenges Index (CRCI), alongside qualitative interviews and focus groups to identify vulnerabilities, gaps, and priorities. This dual approach yields plans that are both data-driven and community-informed, thereby increasing legitimacy and buy-in.
- **Use visuals and plain language in public engagement.** Maps, diagrams, and infographics help residents understand risks and solutions, increasing buy-in.
- **Frame resilience in terms of co-benefits.** Link flood mitigation to economic development, transportation improvements, and public health to gain broader support.

unintentionally excluding the voices of those most affected by hazards and stressors. Vulnerable populations, including low-income households, elderly residents, and individuals without reliable transportation, are frequently left out of conversations shaping land use, infrastructure, and emergency management. Participatory processes can correct this imbalance by ensuring procedural justice (fairness in decision-making), distributive justice (fair allocation of resources), and recognitional justice (acknowledging diverse needs and identities).[4] When communities are actively engaged, resilience efforts reflect their lived experiences, making them more inclusive, practical, and sustainable.

Participatory processes do more than gather input. They reshape priorities, align technical solutions with community needs, and create plans that are both just and adaptive. When community priorities – such as safe evacuation routes, affordable housing, or access to healthcare – are integrated into comprehensive plans or hazard mitigation plans, resilience becomes part of everyday governance.

Integrating resilience and equity across planning instruments

Various planning instruments are available to and used by localities to support resilience efforts. Three common planning instruments are prime opportunities for addressing resilience and equity: (1) comprehensive plans, (2) hazard mitigation plans, and (3) capital improvement plans.

Comprehensive plans

Comprehensive plans, also referred to as general plans, provide the overarching policy framework that guides a community's long-term growth and development, typically over a 10- to 20-year horizon. They address land use, housing, transportation, infrastructure, economic development, natural resources, and public facilities, and influence major local decisions, from zoning updates to capital investments. Because of their broad scope, comprehensive plans are one of the most powerful tools for mainstreaming resilience. Many already touch on resilience indirectly through policies related to floodplain management, infrastructure siting, targeted economic development areas, housing quality, or economic diversification, but these elements are often scattered across chapters rather than treated as a cohesive

Figure 5.2 Integration of resilience and equity across planning instruments.
Source: Created by the authors.

strategy. Our experience with The RAFT shows that integrating resilience into comprehensive plans is most effective when resilience is consistently and strategically embedded throughout.

When comprehensive plans explicitly connect long-term growth decisions to risk reduction and equitable outcomes, they help ensure that resilience becomes a guiding principle for everyday governance rather than an add-on addressed only after disasters. Examples from the Town of Chincoteague, on the Eastern Shore of Virginia, demonstrate how hazard mitigation goals, flood risk considerations, and equity concerns can be woven into land use, housing, and infrastructure policies.

The Town of Chincoteague's 2020 update to the comprehensive plan[5] included eight objectives related to resilience:

1. Develop and implement a proactive plan for the community to defend itself from and adapt to long-term changes in climate and

storm impacts, flooding, water table, vegetation, and other environmental factors.
2. Implement a wastewater management plan through a phased implementation.
3. Implement drainage management strategies to enhance overall drainage capacity.
4. Reduce the vulnerability of evacuation routes and collector roads to stormwater and tidal surge inundation.
5. Perform a comprehensive review of shoreline management to identify information gaps and potential projects to help protect the island on which the Town is located.
6. Encourage increased use of natural and nature-based features to increase coastal resilience.
7. Support the establishment of resiliency goals, strategies, and implementations to assist businesses with resiliency planning.
8. Identify critical infrastructure elements, assess their vulnerability, and enhance their resiliency.

The City of Portsmouth, a RAFT pilot community in the Hampton Roads region of southeastern Virginia, integrated resilience across land use, housing, and economic development sections of its 2018 *Build One Portsmouth* Comprehensive Plan.[6]

Hazard mitigation plans

Hazard mitigation plans (HMPs) are required by the Federal Emergency Management Agency (FEMA) for communities to be eligible for federal mitigation and disaster funding and are among the most common planning documents addressing risk. These plans can provide a strong foundation for resilience planning by identifying hazards, assessing vulnerabilities, and outlining actions to reduce future losses from events such as flooding, hurricanes, and severe storms. Traditionally, HMPs have focused on physical assets and emergency management functions, often with limited attention to social vulnerability or long-term adaptation. When HMPs align with comprehensive plans, land-use policies, and capital improvement programs, they help bridge the gap between disaster-focused planning and long-term resilience-building.

The RAFT experience shows that HMPs can serve as a strong foundation for broader resilience planning when they are not treated

Table 5.2 Ideas for incorporating resilience and equity into comprehensive plans

1. **Embed resilience and equity throughout the plan** – Integrate resilience across different sections of the plan so that resilience and equity touch issues across multiple aspects of comprehensive planning. Consider how socially vulnerable populations and/or underserved communities are impacted across different planning sections to ensure equity is embedded throughout.
2. **Identify and map vulnerabilities** – Include vulnerability and exposure analyses to guide development away from high-risk areas. Analyses could include mapping of flood exposure, shoreline erosion, important community assets, and critical infrastructure such as key transportation corridors. Incorporating social vulnerability mapping helps address equity concerns.
3. **Designate areas for special protection** – Identify and protect critical natural resources and conservation areas such as wetlands and marshes. Also designate key infrastructure, such as main roads, drinking water reservoirs, and major hospitals, as protected areas.
4. **Direct future development away from vulnerable areas** – In future projected land use maps and zoning ordinances, prioritize higher elevation areas for development and discourage development in vulnerable areas such as flood zones.
5. **Establish clear goals, objectives, and policies to address resilience challenges** – Include resilience-specific goals and policies and implementation strategies and link them to measurable outcomes and timelines. For example, the City of Portsmouth's 2018 Comprehensive Plan lists goals and strategies for resilience, with each goal having its own section outlining the tactics to achieve it.
6. **Identify funding or revenue sources for resilience projects** – Outline potential sources of funding for resilience projects, such as tax credits or developers fees, state and federal grants for hazard mitigation, and dedicated taxes.
7. **Create an implementation schedule** – Assign timelines and milestones for resilience actions to ensure accountability. For example, break down implementation into short-term, long-term, and ongoing actions, and pair the project timelines with responsible agencies and funding strategies.

simply as compliance exercises that are developed solely to maintain eligibility for federal aid. The hazard mitigation planning process could include expanded risk assessments to consider climate change and recurrent flooding, and link mitigation actions to equity goals. For example, the HMP can explicitly include social vulnerability mapping, critical infrastructure interdependencies, and strategies for healthcare continuity. These enhancements align with The RAFT's emphasis on equity and systems thinking, ensuring that hazard mitigation planning supports holistic resilience.

Several communities we have worked with have successfully incorporated resilience priorities generated through The RAFT process into their updated HMPs. The RAFT process leverages HMPs as a starting point for identifying possible resilience actions for communities. To help community stakeholders brainstorm ideas for projects that increase resilience, we create an Opportunity List that includes mitigation actions from the community's HMP, ensuring that resilience actions build on existing priorities rather than duplicating efforts. The Middle Peninsula Planning District Commission integrated strategies for recurrent flooding and sea level rise into its updated 2021 HMP.[7] Similarly, the Northern Neck Planning District Commission referenced resilience goals from The RAFT process in its 2023 regional HMP, explicitly connecting hazard mitigation to resilience and equity goals.[8]

HMPs also provide a structured framework for integrating resilience into other plans. Communities can use HMP risk assessments to inform land use decisions, capital improvement prioritization, and zoning updates. For instance, the City of Portsmouth's hazard mitigation plan earned significant FEMA Community Rating System points by mapping repetitive loss areas and outlining policies to address them, which were later incorporated into its comprehensive plan and local ordinances. This integration demonstrates how HMPs can serve as a bridge between emergency management and long-term planning.

Capital improvement plans (CIPs)

Capital improvement plans (CIPs) guide long-term public investments in infrastructure, such as roads, stormwater systems, water and wastewater facilities, public buildings, and parks, typically over a five-year period. While CIPs are often viewed as technical capital budgeting documents, they can play a critical role in determining whether resilience priorities are implemented on the ground.

Incorporating resilience into CIPs means prioritizing projects that reduce vulnerability and enhance adaptive capacity. Many resilience-related projects, including drainage upgrades, facility hardening, and infrastructure relocation, can only occur if they are reflected in capital budgets. The RAFT assessments frequently identify a disconnect between resilience goals and capital spending priorities, particularly in under-resourced communities where funding decisions are tightly constrained. Integrating equity and resilience into the CIP means prioritizing projects that reduce risk in vulnerable neighborhoods, using hazard and climate data to inform project selection, and coordinating capital investments with comprehensive plans and HMPs. When resilience considerations are embedded in capital planning criteria, CIPs can become a practical bridge between long-term vision and near-term action.

Reviews of CIPs in the Historic Triangle region of Virginia (City of Williamsburg, York County, and James City County) identified several stormwater upgrades and wastewater infrastructure as top priorities for resilience investments. James City County, for example, has capital projects related to addressing historical drainage and flooding issues, restoring streams to stabilize degraded channels and reduce bank erosion, and installing a generator at a community center to allow it to function as a general emergency shelter during power outages. These examples show how resilience considerations can shape major infrastructure investments.

Other plans

In addition to the comprehensive, hazard mitigation, and capital improvement plans, other plans, such as land use and transportation plans, can also offer opportunities to embed equity in the planning process and to integrate resilience.

Land use plans

Land use planning translates a community's long-term vision into decisions about where development occurs, at what density, and under what conditions. These decisions fundamentally shape a community's exposure to hazards, the costs of infrastructure, the feasibility and logistics of evacuation, and access to economic activities. In many communities, land-use planning is especially consequential for resilience because it determines whether new development is directed toward less vulnerable areas or continues to concentrate people and assets in hazard-prone locations. Many land use plans already include floodplain restrictions or references to hazard areas, but these measures often focus on minimum regulatory compliance rather than long-term adaptation. When focused on long-term adaptations, land use planning can address resilience by prohibiting construction of new critical facilities (such as hospitals, water treatment plants, key resource distribution centers, etc.) in hazard zones; encouraging compact development on higher ground; requiring open spaces for improving water management; or using rolling easements that allow shorelines to migrate inland over time while preventing the construction of new permanent structures.

Equity challenges can also arise from land use planning when land use policies limit affordable housing options in lower-risk areas or inadvertently reinforce patterns of displacement. Land use planning needs to recognize these equity issues and address them through inclusive planning processes. Land use planning can support equitable resilience by aligning zoning with hazard data, incentivizing adaptive and resilient design, and coordinating land use decisions with housing and infrastructure strategies. When land use planning intentionally reduces risk for all, and not just the advantaged, and balances risk reduction with access to housing, jobs, and services, it can become a central mechanism for shaping a more resilient and equitable future for the community.

Transportation plans

Transportation plans shape how people and goods move through a community on a daily basis and during emergencies. They guide investments in roads, bridges, transit systems, sidewalks, bike infrastructure, and evacuation routes, often at a regional scale. Because transportation systems function as critical lifelines, they are central to community resilience, even when not labeled as such. Flooding, sea level rise, and extreme weather increasingly disrupt roadways and transit systems, affecting access to jobs, healthcare, schools, and emergency services.

Transportation plans already implicitly address resilience through asset management, evacuation planning, and design standards, but they often fall short of identifying who is most affected when systems fail. Qualitative resilience assessments conducted as part of The RAFT application in several regions have surfaced concerns about roadways that flood repeatedly, limited evacuation options for residents without vehicles, and the isolation of rural or low-income neighborhoods during storm events.

Transportation plans can incorporate equity and resilience throughout every facet of the transportation lifecycle. Integrating equity and resilience into transportation planning means using hazard and social vulnerability data together to prioritize investments, coordinating with hazard mitigation and capital improvement planning, and ensuring that evacuation and recovery strategies account for residents with limited mobility or access to private transportation. For example, at the state level, Virginia's statewide transportation plan integrates resilience by requiring risk-based asset assessments and adaptation strategies for sea level rise and storm surge. Several of these resilience strategies can trickle down to the regional and local levels. Transportation projects that address coastal resilience include elevating bridges and using inundation mapping for design modifications. Transportation plans can also incorporate nature-based solutions, such as living shorelines and vegetative buffers, to protect transportation infrastructure while providing ecological benefits.

Sustainability and climate action plans

Sustainability, climate, and resilience are deeply interconnected. A sustainability plan is a comprehensive document that outlines a locality's long-term goals and strategies for environmental, economic, and social well-being. It serves as a roadmap for building and operating a more environmentally sustainable local government. The plan contributes to building a sustainable community and addressing environmental challenges. As such, sustainability plans are typically designed to promote long-term environmental stewardship, efficiency in resource use, and quality of life. They often address energy use, waste reduction, water conservation, green infrastructure, food systems, and environmental health. Many of these strategies can directly contribute to resilience outcomes by reducing exposure to hazards and improving ecosystem services. However, sustainability initiatives can unintentionally exacerbate inequities if benefits such as green investments or energy upgrades accrue primarily to higher-income areas.

Sustainability and climate action plans are closely connected, and while they are sometimes used interchangeably, sustainability plans address broader issues than just climate action. Climate action plans are comprehensive roadmaps that outline specific strategies and activities to reduce greenhouse gas emissions and transition to cleaner energy systems. Many climate action strategies, such as building efficiency improvements and renewable energy deployment, can also reduce vulnerability to climate impacts and increase resilience. However, climate action plans tend to focus primarily on long-term emissions targets which can overlook near-term risks and equity considerations.

In practice, sustainability plans and climate action plans can serve as entry points for resilience discussions, particularly in communities where environmental or climate challenges already resonate. Mainstreaming equitable resilience into sustainability planning is most effective when resilience and equity are made explicit, for example, by prioritizing actions that reduce energy burdens, address heat effects on vulnerable neighborhoods, or improve environmental conditions where residents face cumulative risks such as exposure to air pollutants for residents living near industrial facilities, or exposure to lead from water pipes or paint in older housing. Climate action plans can also incorporate equitable resilience when short-term adaptations and longer-term climate mitigation are addressed together and when equity is embedded in implementation strategies. Aligning climate action plans with hazard mitigation, sustainability, and community resilience can help communities connect global climate goals with local needs and ensure that climate investments also reduce existing vulnerabilities.

Healthcare and public health plans

Healthcare and public health plans focus on emergency preparedness, continuity of care, and coordination among hospitals, clinics, and public health agencies. These plans are critical to resilience because healthcare systems are essential during disasters and recovery periods, particularly for vulnerable populations such as older adults, people with medical conditions, and residents without reliable transportation. Despite their importance, healthcare plans are often developed separately from land use, infrastructure, and hazard mitigation planning. Our experience with The RAFT, particularly working with localities during the COVID-19 pandemic, has identified healthcare planning as an area where integration with resilience is still emerging but becoming increasingly necessary. Disruptions to power, water, transportation, or staffing can quickly undermine healthcare capacity during emergencies and impact important factors such as patient comfort and refrigerated storage for medicines.

Linking hospital preparedness and public health systems to resilience planning will strengthen community capacity to respond to disasters. Integrating resilience considerations into healthcare and public health planning means coordinating them with emergency management, transportation planning, and infrastructure investment decisions, as well as recognizing how social determinants of health[9] also impact resilience. When healthcare systems are connected to broader planning efforts, communities are better positioned to protect public health during both acute disasters and ongoing stressors.

Box 5.2 Social determinants of health

Social determinants of health are the nonmedical conditions in the environments where people are born, live, learn, work, and age that shape their health and quality of life. These include factors like income, education, housing, neighborhood safety, access to nutritious food, and social support. Because these conditions influence opportunities and resources, they play a major role in creating or reducing health inequities.

Resilience zoning

Local ordinances can complement various planning efforts by incentivizing resilience measures through tools like resilience zoning. Resilience-oriented zoning can direct growth away from vulnerable areas and incentivize adaptive measures. The City of Norfolk in the

Five domains of the social determinants of health

❶ Economic Stability
Employment
Food insecurity
Housing instability
Poverty

❷ Education Access and Quality
Early childhood development and education
High school graduation
Enrollment in higher education
Language and literacy

❸ Health Care Access and Quality
Access to health services
Access to primary care
Health literacy

❹ Neighborhood and Built Environment
Housing quality
Transportation
Access to affordable and nutritious foods
Neighborhood safety
Environmental conditions

❺ Social and Community Context
Social support
Community engagement
Discrimination
Civic participation

Figure 5.3 Five domains of the social determinants of health.
Source: Created by the authors.

Hampton Roads region of southeastern Virginia offers a nationally recognized example through its Coastal Resilience Overlay zone. Adopted in 2018 as part of a comprehensive zoning overhaul, the Coastal Resilience Overlay requires new development and redevelopment in flood-prone areas to meet elevated flood resilience standards.[10] These include:

- **Freeboard requirements** – Buildings must be elevated at least 3 feet above the 100-year base flood elevation in high-risk zones, and 18–24 inches in lower-risk areas.
- **Stormwater management** – Sites must retain the first 1.25 inches of rainfall on-site, often through use of rain barrels or permeable surfaces.
- **No new basements** – To reduce flood damage risk, new basements are prohibited citywide.

To complement this, Norfolk also introduced an Upland Resilience Overlay zone, which encourages development in areas with lower flood risk, using the market to help shift growth toward safer terrain.

Additionally, Norfolk's Resilience Quotient is a point-based system embedded in its zoning ordinance that evaluates how well new developments address resilience challenges. In order to be approved, a development proposal must earn a certain amount of points, depending on the size and type of proposed development. It blends required baseline standards with optional enhancements, offering flexibility while promoting best practices. Developers earn points by incorporating features such as:

- **Risk reduction** – Elevating structures, using flood-resistant materials, and designing for wind resilience.
- **Stormwater capture** – Installing rain gardens or green roofs or using tree preservation strategies.
- **Energy resilience** – Adding generator hookups or solar-ready infrastructure.
- **Sustainability incentives** – Achieving LEED Gold certification or equivalent for commercial projects.

This tiered approach, involving required "must do," with optional "should do," and "nice to do" measures, allows Norfolk to embed resilience into everyday development decisions while encouraging innovation and adaptation by developers and property owners.

Similarly, the City of Portsmouth (also in Hampton Roads, Virginia) has taken meaningful steps to integrate resilience into its zoning framework, particularly through updates to its Chesapeake Bay Preservation Area (CBPA) Overlay. Chapter 9.1 of the Portsmouth City Code establishes the CBPA Overlay District, which integrates resilience by requiring developers to account for future climate conditions during the permitting process. In alignment with Virginia's 2020 state mandates, the code embeds specific resiliency criteria that necessitate a resilience assessment for any proposed development in the Resource Protection Area adjacent to perennial water bodies. This assessment must evaluate the impacts of sea level rise, Category 1 storm surge, and coastal flooding over a 30-year period using state-approved climate models. The code explicitly encourages nature-based shoreline stabilization, such as living shorelines, over hard structures like bulkheads, and requires enhanced standards for including and protecting mature trees and native vegetation to maintain natural stormwater buffers.

Water supply plans

Water supply plans ensure that communities have reliable access to safe drinking water now and in the future. These plans address source protection, treatment capacity, infrastructure maintenance, demand projections, and contingency planning for droughts or system failures. In coastal communities, water supply planning increasingly must account for climate-related stressors such as saltwater intrusion into aquifers, flooding of facilities, and variability in precipitation. Water reliability is a resilience concern, being closely tied to public health, emergency response, and economic stability. Equity considerations are also critical, as disruptions to water services or affordability challenges disproportionately affect low-income households, older adults, and rural communities. Integrating resilience into water supply planning requires explicitly considering service continuity and affordability during and after disasters. When water supply planning is connected to broader resilience efforts, it reinforces the idea that access to clean water is foundational to community resilience rather than a standalone infrastructure issue.

Floodplain management plans

Floodplain management plans guide how localities regulate development in flood-prone areas, manage flood risk, and participate in programs such as FEMA's Community Rating System (CRS). These plans typically focus on mapping flood hazards, setting building standards, administering permits, and educating the public about flood insurance. While floodplain management is a core element of hazard mitigation, it often emphasizes regulatory compliance rather than long-term resilience outcomes. Furthermore, without an equity lens, floodplain management policies can restrict housing options, increase costs for lower-income residents, or fail to address flooding that occurs outside mapped flood zones. Floodplain management can play a stronger role in equitable resilience when it is aligned with housing strategies, land use planning, and community engagement. For example, incorporating local knowledge about recurrent flooding through listening sessions with community members and/or service providers coordinating floodplain policies with affordable or accessible housing goals, and using CRS incentives to support outreach in vulnerable neighborhoods can help ensure that floodplain management protects all people and property.

Stormwater management plans

Stormwater management plans address how localities manage rainfall runoff to reduce flooding, improve water quality, and protect infrastructure. As rainfall events become more intense and frequent, stormwater systems are increasingly strained, leading to chronic nuisance flooding even during non-major storm events. Many communities already invest heavily in stormwater infrastructure, but these investments are often driven by regulatory requirements or development pressures rather than equity and resilience considerations. The RAFT qualitative assessments conducted across communities in Virginia consistently identified neighborhood-scale flooding as a major concern, particularly in under-resourced areas with outdated stormwater infrastructure where drainage problems persist for years without resolution. Integrating equity and resilience into stormwater planning could be achieved by using community input to identify priority problem areas, coordinating stormwater upgrades with capital improvement plans, and designing projects that deliver multiple benefits such as green space, canopy and cooling, or improved walkability. When stormwater management is framed as both a technical and social issue, it can become a powerful tool for building trust and reducing everyday disruptions that undermine community resilience.

The Town of Colonial Beach in the Northern Neck region of Virginia combined flood resilience and stormwater management in its planning efforts. The 2024 plan titled "Propelling Resilience" combines infrastructure improvements, nature-based solutions, community engagement, and policy updates to mitigate the impacts of frequent flooding, sea level rise, and climate change.[11] Through the plan, the Town will prioritize natural and nature-based solutions, such as living shorelines and wetland restoration, alongside land acquisition in high-risk areas to absorb floodwaters and reduce wave action. This is further supported by technical upgrades, such as real-time flood monitoring systems, and by administrative actions, including the potential implementation of stormwater fees and the creation of a Resiliency Advisory Committee. By integrating these environmental, financial, and regulatory tools, the Town aims to mitigate the impacts of sea level rise and frequent flooding while ensuring long-term community protection and environmental justice.

Tree canopy and urban forestry plans

Tree canopy and urban forestry plans guide the planting, preservation, and maintenance of trees to support environmental and community benefits. Trees play a significant role in resilience by reducing urban heat, absorbing stormwater, improving air quality, and enhancing neighborhood livability. However, tree canopy is often unevenly distributed, with lower-income neighborhoods and communities of color experiencing higher heat exposure and fewer environmental benefits. Tree canopy and urban forestry plans that do not explicitly address equity risk reinforcing these disparities. Integrating resilience into tree canopy planning involves prioritizing planting and maintenance in heat-vulnerable areas, coordinating with stormwater and green infrastructure strategies, and engaging residents in stewardship and decision-making. Our experience with the application of The RAFT in various Virginia communities suggests that communities are often open to the idea of using trees to enhance community resilience. Tree canopy initiatives can be framed as essential community infrastructure and not just aesthetic improvements as they are often perceived A tree canopy survey and inventory conducted for the Town of Colonial Beach, for example, identified specific areas where tree preservation should be prioritized as a way to manage stormwater runoff and shoreline erosion. Ancient oak trees lining the town's waterfront are considered critical green infrastructure, as their deep roots help to stabilize the sandy shoreline and absorb runoff from frequent storms. Maintaining the canopy in the historic district and near the boardwalk is a resilience focus since the trees protect local businesses from flooding while preserving the community's green and lush character. Expanding resilience more broadly to incorporate tree enthusiasts helps generate broader support for and clearer alignment with resilience goals.

Green infrastructure plans

Green infrastructure plans promote the use of natural and nature-based solutions such as wetlands, living shorelines, bioswales, rain gardens, trees, and permeable surfaces. These approaches help manage flooding, reduce erosion, improve water quality, and enhance ecological resilience. Green infrastructure is often highlighted as a cost-effective and flexible alternative to traditional gray infrastructure (bulkheads, storm drains, revetments, etc.), particularly in coastal settings. However, without intentional planning, green infrastructure investments may be concentrated in highly visible or higher-income areas, limiting their equity benefits. Green infrastructure projects are most effective for integrating equitable resilience when they are designed in response to localized flooding concerns identified by residents and when they deliver multiple co-benefits such as recreational space, cooling, or workforce development opportunities.

Box 5.3 Green and gray infrastructure

Green and gray infrastructure represent complementary approaches to managing water, reducing risk, and supporting community resilience.[12] **Green infrastructure** uses natural or nature-based systems, such as wetlands, green roofs, rain gardens, bioswales, living shorelines, and restored floodplains, to absorb, slow, and filter water while providing co-benefits like habitat, cooling, and improved public space. **Gray infrastructure** relies on engineered, built systems, such as pipes, culverts, storm drains, levees, seawalls, dams, pump stations, and wastewater treatment plants, designed to control, convey, or contain water in predictable ways. When designed together, green infrastructure can reduce the load and extend the lifespan of gray systems (for example, using green streets and wetlands to manage stormwater upstream of pipes and pumps), resulting in more flexible, cost-effective, and equitable infrastructure systems that perform better under climate stress and uncertainty.

Resilience plans

Mainstreaming equitable resilience into existing planning instruments offers a somewhat easier way for communities to build resilience. By embedding equity and resilience across different plans, such as the comprehensive plan and the hazard mitigation plan, communities can take small steps that span different facets of their environmental, economic, and social well-being.

However, some communities are ready and able to tackle resilience through a standalone resilience plan. Such resilience plans are explicitly designed to address how communities prepare for, withstand, and recover from shocks and stressors, including natural hazards, economic disruptions, and social challenges. A resilience plan provides an explicit statement of the community's resilience goals, a shared framework for action, and clarification of priorities. The plan facilitates project implementation and potentially supports funding applications.

However, a standalone resilience plan can risk becoming disconnected from day-to-day decision-making if it is not integrated with other planning processes. In this sense, resilience plans also build upon the various planning tools and documents that already exist. Some localities have resilience plans that are developed as a standalone document. For example, the City of Hampton, in the Hampton Roads region of southeastern Virginia, developed a standalone resilience plan titled "Living with Water Hampton: A Holistic Approach to Addressing Sea Level Rise and Resiliency" that includes four categories of action: (1) policy actions, such as land use, zoning, and building codes; (2) education and communication around public information, community engagement, and emergency communications; (3) physical or structural projects encompassing gray and green solutions; and (4) operations and maintenance actions to ensure long-term performance of infrastructure and natural systems.[13] The City's resilience plan also incorporates neighborhood-specific strategies that enable actions tailored to the hydrology and geology of the different neighborhoods across the City.

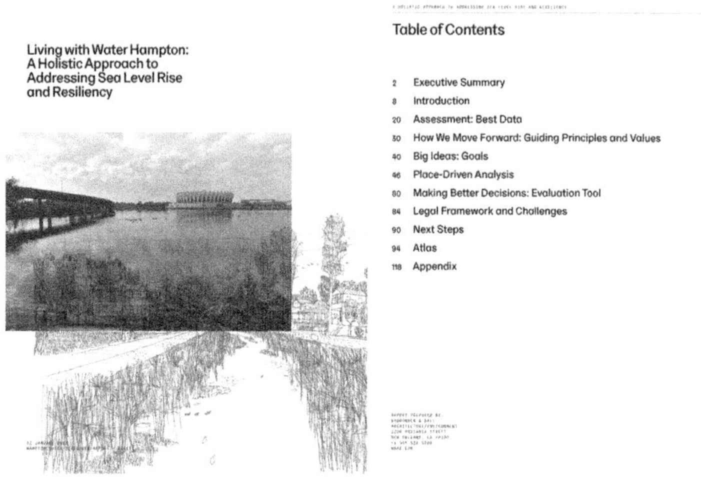

Figure 5.4 The City of Hampton's resilience plan.

Source: City of Hampton, Community Development. (2018). *Living with Water Hampton: A Holistic Approach to Addressing Sea Level Rise and Resiliency.* Cover page and table of contents shown. Publicly available at: www.hampton.gov/DocumentCenter/View/20644/Resilient-Hampton-Phase-I-Report.

Other localities approach resilience plans as coordination documents. Our experience with The RAFT suggests that resilience plans are most effective when they build on and draw directly from existing plans, including comprehensive, hazard mitigation, capital improvement, and sector-specific plans. Rather than replacing these efforts, a resilience plan can serve as a coordinating document that aligns goals, elevates equity considerations, and translates community priorities into actionable steps. In this way, resilience planning becomes a unifying process that changes how decisions are made across the community.

For example, when working with the Northern Neck Planning District Commission (the regional planning organization for the Northern Neck region of Virginia) to develop the regional resilience plan, we drew from various other processes and plans, including The RAFT Scorecard, the regional hazard mitigation plan, the comprehensive economic development strategy, and the region's water supply plan. As shown in Figure 5.5, the Northern Neck resilience plan included elements from different plans.

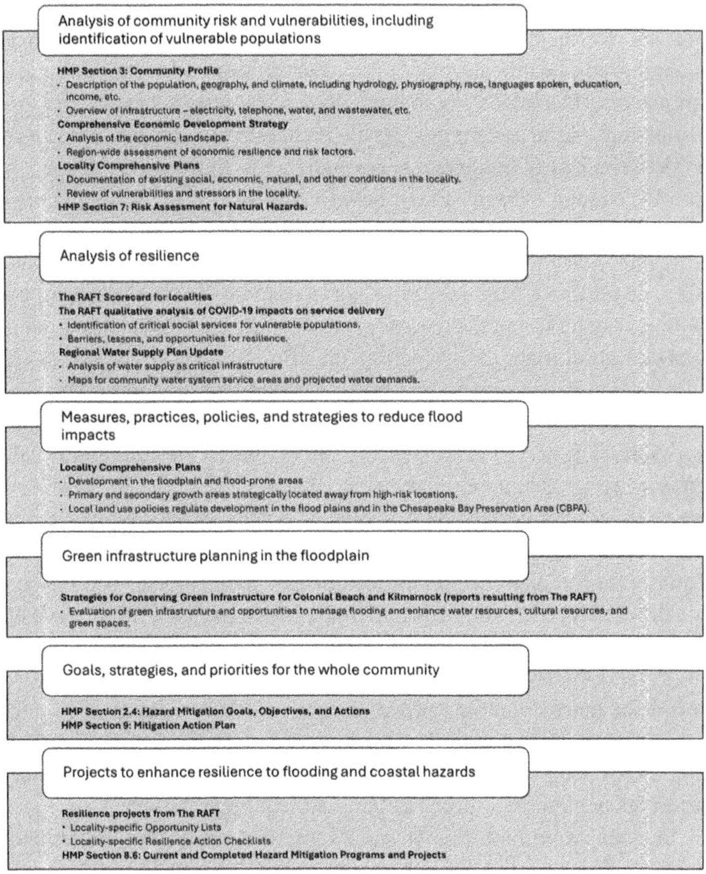

Figure 5.5 The Northern Neck regional resilience plan is a coordination document that draws from other plans.

Source: Created by the authors.

Reflections from The RAFT

Mainstreaming resilience into existing planning processes is cost-effective, equity-driven, and essential for long-term community viability. The RAFT process – from resilience assessment to project identification and implementation – demonstrates how a structured, participatory approach can connect plans, build capacity, and position communities for future funding and success. By embedding resilience across comprehensive plans, hazard mitigation strategies, capital improvement programs, and other areas such as transportation planning and healthcare planning, communities can create integrated frameworks that address both immediate hazards and chronic stressors. Ultimately, mainstreaming equitable resilience ensures that it becomes a core principle of governance, advancing equity for all residents.

This chapter's discussion of how to consider equity in the planning process and embed resilience into these different planning instruments illustrates that equitable resilience is not the responsibility of any single plan, department, function, or sector. Comprehensive plans, land use policies, capital improvement programs, hazard mitigation plans, transportation plans, and function-specific efforts such as healthcare, stormwater, water supply, and green infrastructure planning all shape a community's ability to withstand and recover from shocks and stressors. The RAFT experience demonstrates that equitable resilience is most effectively advanced when these existing plans are intentionally connected via a resilience focus, rather than treated as isolated or purely technical documents. By embedding resilience and equity into the plans communities already update, fund, and rely on, localities can reduce duplication, make better use of limited capacity, and ensure that efforts to reduce risk also expand access to opportunity, and increase community health and wellbeing. Mainstreaming resilience across planning processes shifts it from a reactive, hazard-focused activity to a shared framework for everyday decision-making that aligns long-term vision, near-term investments, and community priorities to support a more resilient and equitable future.

Several important points have emerged from our experience applying The RAFT in different communities. First, resilience planning must adopt a multi-dimensional approach that includes social, economic, and environmental considerations. Second, systems thinking helps overcome policy silos and create multifunctional strategies that integrate disaster risk reduction, infrastructure improvements, and land-use planning. Third, stakeholder engagement is essential. Trusted messengers can help reach vulnerable populations, and co-production processes foster social learning and durable solutions. Finally, equity must be operationalized through targeted actions and resources for vulnerable populations and by incorporating grassroots priorities alongside technical assessments.

Lessons learned for practice

1) **Mainstreaming resilience leverages existing mandates and plans** – Communities already update comprehensive plans every five to ten years and hazard mitigation plans every five years for compliance with federal or state requirements. Resilience can be embedded into these plans and planning processes without creating new planning burdens, thus saving time and resources. By embedding resilience and equity into the plans communities already update, fund, and rely on, localities can reduce duplication, make better use of limited capacity, and ensure that efforts to reduce risk also expand access to opportunity, and increase community health and wellbeing.
2) **Equity must be operationalized, not just stated** – The RAFT's qualitative assessments and whole-community approach ensure that the needs of vulnerable populations are represented in the process of setting resilience priorities. This builds procedural and distributive justice into resilience planning. For example, The RAFT community workshops and implementation teams often include social service agencies and grassroots leaders to ensure equity concerns are reflected in prioritized actions.
3) **A resilience plan should draw from and build on existing plans** – Communities with resources and capacity can create standalone resilience plans that serve as an integrative document that explicitly addresses resilience. However, a resilience plan can also draw from existing plans and still serve as an integrative document that centers resilience. This approach allows under-resourced localities to create and have a resilience plan to guide decision-making across the whole community. For example, the outcomes of a resilience assessment, along with sections of the hazard mitigation plan that address risk and vulnerability, and vulnerable populations specifically, can be referenced in the resilience plan as part of establishing the community's resilience profile. Elements from the transportation plan and water supply plan can provide an assessment of infrastructure vulnerability and guidance for investment in infrastructure resilience projects. Strategies and actions from the comprehensive plan and hazard mitigation plan provide a list of prioritized actions for building resilience.

Notes

1 Braunschweiger, D., & Pütz, M. (2021). Climate adaptation in practice: How mainstreaming strategies matter for policy integration. *Environmental Policy and Governance*, *31*, 361–373.
2 Kok, M., & De Coninck, H. (2007). Widening the scope of policies to address climate change: Directions for mainstreaming. *Environmental Science & Policy*, *10*(7), 587–599.
3 Wu, J., & Thomann, E. (2023). Governance in public policy. In M. van Gerven, C. Rothmayr Allison, & K. Schubert (Eds.), *Encyclopedia of Public Policy* (pp. 1–11). Springer International Publishing.
4 Fox, N., Tilt, J. H., Ruggiero, P., Stanton, K., & Bolte, J. (2023). Toward equitable coastal community resilience: Incorporating principles of equity and justice in coastal hazard adaptation. *Cambridge Prisms: Coastal Futures*, *1*, e36.
5 Town of Chincoteague. (2020). *Comprehensive Plan: 2020 Update*. Chincoteague, VA.
6 City of Portsmouth. (2018). *Build One Portsmouth – Portsmouth 2018 Comprehensive Plan*. Portsmouth, VA.
7 Middle Peninsula Planning District Commission. (2021). *Middle Peninsula All Hazards Mitigation Plan*. Middle Peninsula Planning District Commission, VA.
8 Northern Neck Planning District Commission. (2023). *Regional Hazard Mitigation Plan 2023 Update*. Northern Neck Planning District Commission, VA.
9 Marmot, M. (2005). Social determinants of health inequalities. *The Lancet*, *365*(9464), 1099–1104. https://doi.org/10.1016/S0140-6736(05)71146-6
10 Pew Charitable Trusts. (2019). *Norfolk's Revised Zoning Ordinance Aims to Improve Flood Resilience* (Mitigation Matters: Policy Solutions to Reduce Local Flood Risk). The Pew Charitable Trusts.
11 Town of Colonial Beach. (2024). *Propelling Resilience: Flood Resilience and Stormwater Management Plan*. Colonial Beach, VA.
12 Browder, G., Ozment, S., Rehberger Bescos, I., Gartner, T., & Lange, G.-M. (2019). *Integrating Green and Gray: Creating Next Generation Infrastructure*. World Bank & World Resources Institute.
13 City of Hampton. (2018). *Living with Water Hampton: A Holistic Approach to Addressing Sea Level Rise and Resiliency*. Hampton, VA.

6 Growing and sustaining whole-community resilience for the long haul

Introduction

The common thread throughout our work in Virginia's communities has been a whole-community approach, aspiring to bring together local government with community services and health providers, faith and civic leaders, and schools and youth programs. This aspiration has been one of the most challenging aspects of The Resilience Adaptation Feasibility Tool (RAFT), and an area of continual learning and adaptation.

The RAFT's whole-community approach pairs a quantitative scorecard that assesses leadership, infrastructure, economic health, vulnerable populations, and health and well-being with qualitative engagement that brings community-facing sectors into shared decision-making. To prioritize resilience actions for implementation, we invite a wide range of stakeholders, and over time, we have been successful in bringing many voices, concerns, and interests to the table. While there continues to be space for growing greater diversity and inclusion in engagement, our approach has matured significantly in the last decade. For example, in the Town of Warsaw in the Northern Neck of Virginia, the Implementation Team emphasized the need to strengthen the resilience of small businesses in the region. Small business proprietors and the downtown business association actively supported the development of a small business resilience assessment based on the challenges they faced. In the Middle Peninsula region, various nonprofits were engaged in implementation projects around community gardens and food insecurity, connecting this issue to the region's broader community resilience.

DOI: 10.4324/9781003596233-6

However, remembering the old adage of the hammer – if you have a hammer, everything looks like a nail – it's important to realize that just because one whole community approach works in Community A, it may be completely ineffective in Community B. This was a hard lesson to learn, as success in one place naturally builds confidence that the same approach will work elsewhere. Yet, working across five regions in Virginia and with different Virginia Indian Tribes, The RAFT's experience underscores the importance of adapting and tailoring the whole-community engagement approach to the unique culture and needs of each community.

Connecting the dots across the whole community

Achieving equitable climate resilience requires more than a collection of isolated projects or well-intentioned plans. It relies on the ability of communities to connect the dots across the many systems, sectors, and stakeholders whose decisions shape resilience outcomes. A community's resilience is inherently interconnected: flooding affects transportation and evacuation routes; displacement pressures housing; climate-driven disruptions strain food systems, health services, communications, and public safety. For resilience to be equitable, these functional areas cannot be addressed in silos. Instead, they must be understood as interdependent components of a broader community infrastructure that determines who thrives, who struggles, and who is left at disproportionate risk.

Coordinated partnerships are essential to sustaining progress. No single entity, government department, nonprofit organization, private business, or community group, has the authority, resources, or perspective to address all resilience challenges alone. Local government planning departments may manage land use and infrastructure, but community health providers understand the needs of medically fragile residents; schools and youth programs identify barriers faced by families; food pantries and farmers recognize vulnerabilities in food access; emergency managers understand gaps in preparedness and communication. Connecting these insights creates a more complete and accurate understanding of risk and capacity across a community.

For equitable resilience, coordination must extend across both functions and sectors. Functionally, this means ensuring that transportation plans account for the mobility needs of low-income residents during storms; that housing strategies anticipate climate impacts and prevent displacement; that health systems are prepared for climate-exacerbated conditions; and that communications networks reach residents who have historically been excluded from information flows. When these systems are aligned, resilience efforts reinforce rather than undermine one another.

Sector-to-sector coordination is equally important. Government agencies hold decision-making power and control regulatory frameworks, but their effectiveness increases when supported by non-profit organizations with deep relationships in communities, private sector partners with technical expertise and resources, and community leaders who bring lived experience and trusted perspectives. A whole-community approach works only when each sector acknowledges the roles and capacities of others, and when processes make space for meaningful collaboration.

The RAFT's decade of work with communities illustrates both the opportunities and challenges of this cross-sector coordination. In some communities, the coalition of the willing – those who have time, interest, and capacity – becomes an initial foundation for progress. Over time, as trust builds and communication networks expand, additional stakeholders, including small businesses, health providers, faith leaders, and school personnel, can be intentionally brought into the effort. In other communities, deeply rooted silos or strained local capacity require creative solutions, such as paid liaisons, community builders, or partnerships with regional organizations to support coordination.

Connecting the dots also requires intentional attention to equity. Historically overlooked groups, such as renters, mobile home residents, undocumented workers, aging and disabled adults, and under-resourced families, must see their priorities reflected in resilience actions. Their participation often depends on reducing barriers such as transportation, time, compensation, or language access. When these voices shape decision-making, resilience strategies become more culturally relevant, more widely supported, and more effective in reducing harm.

Ultimately, equitable resilience is not a destination but a system of relationships across plans, across sectors, and across communities. Connecting the dots ensures that resilience is shared, coordinated, and sustained over time, advancing well-being for all.

Working with a coalition of the willing within available resources and time

In The RAFT's first engagement with coastal communities on the Eastern Shore of Virginia, we struggled for a full year to bring leaders

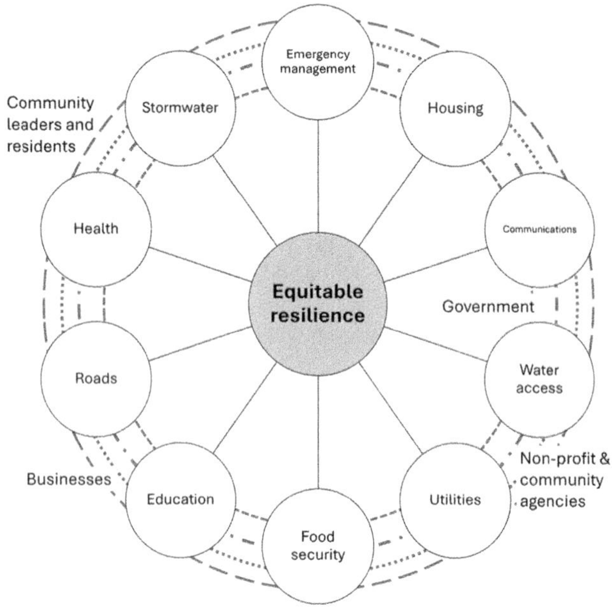

Figure 6.1 Equitable resilience spans functions and sectors.
Source: Created by the authors.

from key parts of the community to the table in all seven communities. This was despite months of outreach via email and phone calls to faith leaders, agricultural leaders, and watermen throughout the year of implementation. As a result, it became clear that The RAFT resilience-building process would need to work with a coalition of the willing, meaning those who have the time, interest, and ability to participate. Community engagement is a messy, imperfect art that requires adapting to what is possible at any given time.

In other communities, we were challenged in getting locality staff to participate in implementing resilience projects. Local government staff might be active participants early in the resilience planning process, such as during the resilience assessment and the priority-setting community workshop, but then become too busy to participate or even to respond to emails and phone calls. When faced with these scenarios, we recruited an Implementation Team of civic and nonprofit and business leaders to rally their human capacity and resources around implementing resilience projects they were able to support. In situations where local government staff lack capacity or interest in implementing resilience projects, working with a coalition of the willing, intentionally designed to include many different civic and community leaders, staff of nonprofit organizations, business leaders, faith-based leaders, and others can potentially achieve the community's resilience goals or at least significantly advance them. However, it is also important to recognize that inevitably there will be community members who are unable to or may choose not to participate because they have children and childcare responsibilities, work full-time jobs or multiple jobs, commute long distances and rely on public transportation, or have no extra capacity to volunteer on a resilience planning committee or implementation team. In this instance, organizations and advocacy groups that represent these populations can be a productive mechanism for accessing the interests and needs of the wider community.

Four years after beginning RAFT implementation on Virginia's Eastern Shore, with climate impacts driving greater interest in building long-term resilience, another project funded by the National Science Foundation proved more successful for a whole-community approach. Building on the earlier application of The RAFT, this whole-community approach involved identifying and paying an annual stipend of $5,000 to two interested and willing community leaders to serve as the project's community liaisons, and $500 honoraria to ten others representing a broad range of interests who would be willing to serve on a Community Advisory Committee. This approach eventually was successful at engaging leaders of faith, local schools, youth, health services, and more. However, it required not only the funding to enable community leaders to justify taking time away from their primary duties to participate in the resilience effort, but it also took several years of effort building on earlier work in the region to accomplish this deeper and broader level of engagement. In other words, achieving a whole-community approach to building resilience may require resources and time that are not yet available. In this case, being open to working with a coalition of the willing and leveraging this early effort to pursue deeper and wider engagement for later or subsequent efforts yielded positive outcomes. Such an approach may be useful in other contexts, too, especially if you proactively try to bring more diversity and new perspectives in throughout the implementation phase of resilience projects.

Figure 6.2 A community stakeholder identifying resilience challenges.
Source: Photograph provided by Virginia Sea Grant. Used with permission.

Moreover, proactively networking with community foundations or other local funders may lead to additional resources that could enhance your engagement process and allow you to be more inclusive by offering honoraria or stipends. If the resources and time are available, you will have a greater chance of successfully creating cross-community discussions about resilience and cross-sector networks that will build the community's resilience to acute and chronic stressors.

Broadening the reach of existing programs

In some localities, The RAFT's whole-community approach calls for expanding or broadening existing programs or services to better connect different sectors and functions. For example, in one coastal Virginia region, one of the top community resilience priorities identified during The RAFT community workshops was to establish a multi-sector, cross-silo environmental justice group that would continue beyond The RAFT's term to work on resilience issues related to health and well-being for lower-income families. Through our ongoing networking and connections with various groups and individuals, it took over six months to learn that such a group already existed, unbeknownst to the regional planning organization that was working with us. With The RAFT, our goal was not to recreate the wheel and unnecessarily duplicate efforts. Instead, we worked with the existing group, which had been meeting monthly for years, to introduce them to the value of connecting health care issues with environmental and other social needs to build community resilience.

Challenges for ongoing community resilience building

Most communities that we have worked with are eager to ensure that the work of The RAFT can continue beyond The RAFT's 18-month engagement process. The RAFT could amplify its impact manyfold by helping each community create an ongoing collaborative workgroup dedicated to identifying annual resilience priorities and working to implement those priorities. Yet achieving this goal remains one of the greatest challenges.

In one region, a pre-existing, region-wide climate-oriented working group agreed to continue the role of convening cross-sector leaders to address resilience issues. But over the next few years, we learned that this working group had not included or invited representatives from the towns in the region, who consequently felt left out. The working group also did not meet regularly, and was not oriented to specific, concrete resilience project implementation in the same way The RAFT had been. As a result, this approach of relying on an existing group with its own agenda and way of doing things was not successful in sustaining the whole-community resilience building efforts of The RAFT.

The basic idea: Resilience committees

During The RAFT implementation, we recommend that communities prioritize establishing a long-term resilience committee as a project on their Resilience Action Checklist, as a way to continue coordinating the strategic advancement of their resilience priorities. We created a worksheet with step-by-step guidance for creating a resilience committee to assist localities that had identified long-term resilience planning as a priority on their Resilience Action Checklists. Figure 6.3 summarizes the three steps for creating a resilience committee from the worksheet.

Forming a resilience committee begins with identifying the specific resilience action or priority that will ground the group's purpose. Stakeholders then collectively define the committee's role, whether advisory, project-focused, or community-facing, and determine the scope of issues it will address, such as emergency management, infrastructure resilience, policy development, or community engagement. This early framing ensures the committee is aligned with community goals and has a clear mandate before membership and structure are developed.

Once the purpose and scope are defined, stakeholders design the committee's formation process, membership composition, and operating structure. This includes deciding how members will be selected and ensuring representation from those most affected by resilience challenges, as well as individuals with technical expertise and community insight. The group then outlines meeting frequency, expected time commitments, leadership roles (such as chair, co-chair, or

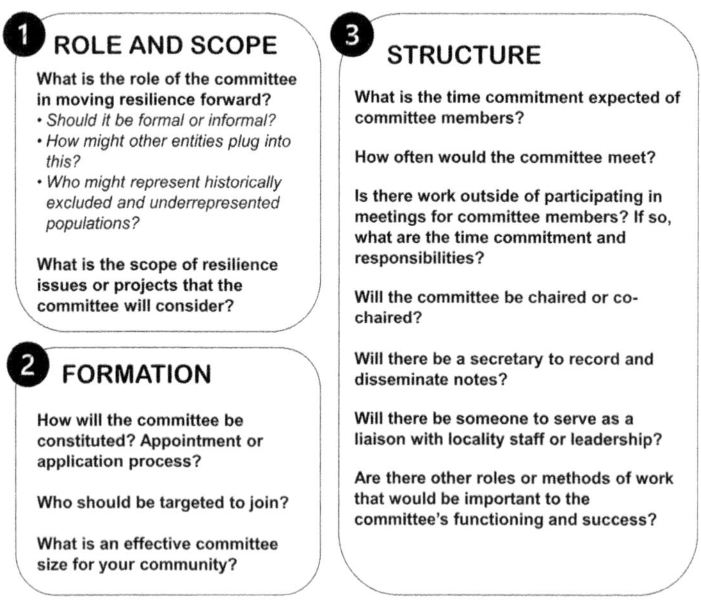

Figure 6.3 Steps to creating a resilience committee.

Source: Created by the authors.

secretary), and any additional working groups or liaison positions needed for effective functioning. Finally, decisions are summarized in an emerging framework that documents the committee's role, scope, formation approach, membership, and leadership structure.

We used the worksheet to assist the Town of Colonial Beach and the Counties of Richmond, Westmoreland, and Lancaster in the Northern Neck region of Virginia in developing a resilience committee structure and plan customized to their unique needs through a series of facilitated workshops. While each locality developed a plan and structure, including clear next steps for launching their committee, only Colonial Beach followed through with formally establishing their resilience committee. This committee met for several months but ultimately dissolved after Town staff decided to prioritize their resilience planning efforts in a more targeted way by developing a Resilience Plan supported by state funding through the Virginia Community Flood Preparedness Funds.

For the three counties that participated in the workshops, staff departures and capacity limitations of county staff to even convene an initial meeting, let alone facilitate these committees in the long term, prevented any progress. Despite clear identification of who would fill which role on the resilience committee, the lack of organizational capacity to coordinate, schedule, plan, and facilitate meetings turned out to be the weak link for each community. Based on subsequent conversations with staff of the regional planning organization and the localities, we identified the main barriers as the competing priorities of county staff and the limited time to facilitate community-engaged resilience committees beyond their core job descriptions. Through this feedback, we learned that we greatly underestimated the capacity limitations of localities to sustain work beyond our team's engagement through The RAFT, and realized that to be successful in the future we would need to allocate more resources and planning to such efforts to sustain resilience efforts in the long term.

While a number of RAFT alumni localities have opted into this recommendation to establish a long-term focus on resilience and even used The RAFT's worksheet and facilitated support to develop a plan for establishing and sustaining a resilience committee, most efforts have been unsuccessful in maintaining these committees long-term. One of the greatest challenges that The RAFT communities and localities share with us in terms of maintaining coordinated work on resilience is the lack of organizational capacity to schedule, plan, and facilitate meetings needed to continue to coordinate resilience efforts. Resilience committee members – all community leaders from community or nonprofit organizations, businesses, service providers, and local government – were working other jobs, often already wearing multiple hats in their communities. Though they wanted to see it succeed, they just did not have the time and capacity to take on this additional responsibility.

The RAFT team explored numerous options for addressing this challenge of community organizational capacity. One option was to offer an honorarium or a stipend to a community person to take on the coordination and facilitation role. There were ample interest and enthusiasm among resilience committee members to continue the resilience collaboration. However, most responded that they already filled two or three other roles in their community. An honorarium or stipend would not create more time in the day for them to invest in this initiative. This frequent response from The RAFT Implementation Team members – a willingness to keep serving on a long-term resilience committee, but no capacity to plan and organize it – reaffirms the somewhat surprising importance of the administrative support and facilitation provided by The RAFT. That is an important lesson that can help any community wishing to undertake long-term resilience efforts: time is a precious resource. It takes time to get into the grassroots level of a community; time to gather community representatives' input, earn their trust, and obtain their participation in project implementation efforts; and it takes time to sustain resilience efforts long-term.

An attempt at a regional solution: Regional Resilience Equity Workgroup

In 2022 and 2023, with support from the Jessie Ball duPont Fund, we facilitated the creation of a regional social equity workgroup to continue advancing resilience priorities that serve the 10-county Northern Neck and Middle Peninsula regions of Virginia. The goal was to create a mechanism to sustain equitable resilience actions following the success of The RAFT and the micro and mini grant program. The group that was established was called the Regional Resilience Equity Workgroup (RREW).

We facilitated two planning workgroup sessions to develop a structure and timeline for the RREW. Honoraria were offered to formal liaisons with four Virginia Indian Tribes in the region as well as to community leaders and nonprofits to allow for their participation.

In January 2023, supported by The RAFT facilitators, the RREW convened its first core workgroup meeting and established subgroups focused on six areas of critical need: (1) economic development and living wages, (2) health and mental well-being, (3) affordable housing and shelters, (4) transit, (5) food security, and (6) programs for youth and families. As articulated by RREW members, the purpose of this coalition was to "create a space where groups can collaborate and share information on resilience for the underserved and vulnerable, for the specific purpose of identifying and resolving gaps."

The RREW functioned as a coalition of community organizations, nonprofits, local and state agencies, and Virginia Indian Tribes dedicated to advancing equitable solutions to resilience challenges faced by the ten counties. The RAFT team provided a critical coordination role for the workgroup, facilitating monthly RREW Core Workgroup meetings, and providing support to each of the six subgroups' leads and co-leads from January to December 2023.

Recognizing that organizational capacity would continue to be a challenge for the RREW's long-term sustainability, we partnered with Bay Aging, a regional nonprofit organization, to apply for and recruit an AmeriCorps VISTA Capacity-Building position to provide critical coordination support for the RREW after our facilitation ended. While this was an effective model to sustain the RREW initially, Bay Aging and the RREW faced difficulty recruiting a second AmeriCorps VISTA position which, when coupled with the federal funding gaps in 2025, meant this model would no longer be feasible as an affordable way to provide additional capacity support for coordinating a long-term resilience committee.

We sought to sustain the RREW through an AmeriCorps VISTA position that would add the organizational capacity needed to provide vital organizational coordination support and community capacity building, coupled with providing each member of the workgroup with honoraria or stipend funding for their participation. Thus, none of the already overcommitted community members would need to take on the role of coordinator and facilitator. Additionally, they would be able to justify their participation to their parent organization by virtue of the honoraria. This AmeriCorps VISTA recruit *plus* participant honoraria approach was partially successful, for the short-term. The downside of this approach was that the VISTA recruits are often recent college graduates with limited work experience and not typically recruited from the local community, thus requiring significant mentoring from The RAFT team. Additionally, the placement lasts for only one year, new recruitment and training must occur each year under that approach.

Building capacity through community builders and liaisons: A shared framework for local leadership

While the RREW is The RAFT's greatest success to-date in terms of sustaining a diverse coalition of organizations and institutions working on whole-community resilience issues beyond The RAFT process, its future as a long-term mechanism is still emerging. Through the work of establishing and supporting the RREW, we identified an important lesson: long-term, sustained collaboration between localities and community-based organizations around resilience planning requires significant capacity-building and dedicated funding to support the ongoing work of a facilitator or coordinator. Localities can only remain engaged in resilience planning, particularly in ways that genuinely incorporate community perspectives, if there is additional, dedicated facilitation and coordination capacity beyond what locality staff can sustain on their own.

Across our work in Virginia, successful application of The RAFT has relied on compensated, community-rooted leaders who strengthen trust, elevate local priorities, and ensure that resilience planning and implementation reflect lived experience and local governance structures. These positions, in the form of Community Builders, Community Network Builders, or Community or Tribal Liaisons, support a shared model for building durable community capacity.

The RAFT has actively resourced these roles in multiple regions. In Southwest Virginia, a Community Builder hired from within the community now works 32 hours per week conducting direct engagement and relationship-building in two rural counties. In subsequent work on the Eastern Shore, the Northern Neck, and the Middle Peninsula, similar community-based roles have been supported to ensure long-term engagement. On the Eastern Shore, a Community Advisory Committee, supported through a project funded by the National Science Foundation Coastlines and People (NSF CoPe) program, formalizes this approach through paid Community Co-Leads and a 12-member leadership structure representing key organizations in the region. In the Northern Neck and Middle Peninsula regions of Virginia, the Jessie Ball duPont Fund is supporting a full-time

Community Network Builder, hired by the University of Virginia but recruited from the local area, to coordinate regional collaboration, convene partners, and support cross-jurisdictional resilience and economic development initiatives.

Compensation for this whole-community resilience-related work is important. These roles require time, relational labor, and community expertise. Treating them as paid positions ensures continuity, accountability, and sustained partnership, which are essential for continued resilience implementation beyond The RAFT.

For The RAFT's work with tribal nations, this community builder model has been adapted into the role of a Tribal Liaison who serves as the Tribe's formally designated point of contact for The RAFT team, appointed by the Chief or Tribal Council. The Tribal Liaison ensures that all resilience activities undertaken as part of The RAFT respect tribal sovereignty, governance protocols, and leadership priorities. The liaison responsibilities include reviewing tribal policies, scheduling and coordinating engagement activities, guiding adaptation of The RAFT Scorecard to the tribe's context, and supporting tribal participation throughout the assessment, workshop, and implementation phases. Tribal Liaisons are compensated to recognize the significant time and expertise required to engage tribal staff, Council members, elders, and community members. This structure prevents additional strain on already limited tribal staff capacity and affirms that cultural knowledge, governance insight, and coordination work are forms of professional expertise. It also supports data sovereignty, keeping control of information and decision-making within the tribe.

Working across Southwest Virginia, the Eastern Shore, the Northern Neck and Middle Peninsula, and with tribal nations in Virginia, The RAFT's experience demonstrates that community-driven resilience requires community-embedded, compensated leadership. Whether called a Community Builder, Community Network Builder, or Community/Tribal Liaison, these roles form the essential infrastructure for:

- Sustained collaboration after The RAFT's formal processes end.
- Trust-based engagement with residents and institutions.
- Cross-sector and cross-jurisdictional coordination.
- Culturally grounded, community-owned resilience planning.
- Implementation capacity through mini-grants, tool development, and structured facilitation.

Ultimately, without these compensated roles, the long-term, community-centered resilience work sparked by The RAFT cannot be sustained. These positions are the backbone of durable, community-driven resilience planning.

Lessons learned for practice

- **Customize whole-community engagement to fit the local context, not a fixed model** – A whole-community approach is not one-size-fits-all. What worked in one coastal town or region may fail elsewhere due to differences in culture, leadership structures, history, and trust. We see The RAFT engagement strategies as adaptable frameworks rather than replicable templates. This is especially important for equitable resilience, where cultural relevance and local legitimacy determine who participates and whose priorities are elevated.
- **Be realistic about local government capacity and plan alternative leadership pathways** – Local government staff often support resilience planning in theory but lack the time or mandate to lead long-term, community-engaged processes. Practitioners should avoid assuming that localities can absorb additional coordination responsibilities and instead plan for shared or non-governmental leadership models that include nonprofits, tribes, regional organizations, or paid community builders. This flexibility is critical for advancing equity when formal institutions are overstretched. Where possible, clarify who owns the coordination role after the resilience process ends and how that coordination work will be funded.
- **Start with a coalition of the willing, and plan intentionally for broader inclusion over time** – In many communities, only a subset of leaders or organizations have the time or capacity to engage initially. Working with a coalition of the willing can create early momentum, demonstrate value, and deliver near-term outcomes. However, equitable resilience planning requires a clear strategy for expanding participation later to include underrepresented groups, sectors, and voices once trust, resources, or capacity increase. Identify which voices are missing and set a timeline and approach (e.g., offering stipends, providing childcare, improving language access) to include them.

- **Recognize that participation has real costs and allocate resources accordingly** – Expecting sustained engagement without compensation disproportionately excludes lower-income residents, nonprofit staff, tribal representatives, and community leaders who already carry multiple responsibilities. Budget for honoraria/stipends, paid liaison roles, and practical supports (childcare, transportation, interpretation) from the outset. Treat compensation not as an incentive but as a core equity practice.
- **Invest time in broadening or expanding existing networks and initiatives** – Well-established community groups, coalitions, or programs may already be addressing resilience-related issues such as disaster preparedness, health, food security, or environmental justice. Failing to identify these can lead to duplication, frustration, and missed opportunities. Conduct a quick landscape scan and meet groups where they are, prioritizing collaboration over creating new structures unless a clear gap exists.
- **Treat coordination and facilitation as core infrastructure, not an add-on** – The most consistent barrier to sustaining resilience committees and workgroups has been not a lack of interest, but lack of organizational capacity to coordinate meetings, follow up, and maintain momentum. Equitable resilience planning depends on dedicated facilitation and coordination roles, whether housed in local government, a nonprofit partner, or a regional entity. Without this infrastructure, even well-designed collaborative efforts are unlikely to last.
- **Pair long-term resilience goals with sustained investment in people, not just plans** – Resilience plans, committees, or workgroups cannot be sustained through short-term funding and projects alone. Our experience shows that durable, equitable resilience requires ongoing investment in people who can convene, translate, and connect across sectors. It is important to advocate for and pursue funding models that prioritize the long-term capacity-building of locally embedded people and organizations over short-term deliverables.

7 Partnering for resilience

Guidance for effective partnerships

Introduction

Federal and state agencies, nonprofit organizations, extension services, regional planning bodies, and other boundary-spanning organizations have long played important roles in providing technical assistance, funding, facilitation, and coordination. Local governments have experience working with at least some of these partners, whether through hazard mitigation planning, coastal zone management, public health initiatives, or disaster recovery programs. These partnerships are essential and often form the backbone of resilience work.

Less frequently recognized, however, are colleges and universities as potential partners in equitable resilience efforts. Institutions of higher education are often physically located in or near the very communities grappling with resilience challenges. They house disciplinary expertise across the natural sciences, engineering, social sciences, planning, law and policy, public health, and the humanities. They also have a steady flow of students seeking applied learning opportunities and faculty who are encouraged, and in some cases required, to engage in research, teaching, and service that benefits society. Despite this alignment of needs and capacities, universities are often overlooked as resources in community-based resilience planning or are engaged only episodically and without long-term coordination.

We believe that colleges and universities represent a significant but underutilized resource for resilience planning and implementation, particularly for under-resourced communities. At the same time, our experiences have identified a core lesson: universities often need help in providing direct support to communities. Academic institutions are not naturally structured to respond to

DOI: 10.4324/9781003596233-7

community needs on timelines appropriate for local governments and tribes, nor are communities typically resourced to manage academic partnerships on their own.[1]

Drawing on our experiences with The Resilience Adaptation Feasibility Tool (RAFT), this chapter explores how colleges and universities can play a more intentional and effective role in supporting coastal community resilience and discusses both the opportunities and the challenges of university-community collaboration. In particular, it focuses on work with under-resourced communities, where capacity constraints make external partnerships especially important but also more complex.

Through examples of class projects, internships, applied research, and service-learning efforts (some examples are offered in Chapter 4), we illustrate how students, faculty, and staff can contribute meaningfully to resilience assessments, project identification and implementation, and planning processes. At the same time, this chapter highlights the structural barriers that can limit the effectiveness of university engagement, including mismatches in timelines, incentives, communication styles, and expectations. It examines common engagement models, barriers to collaboration, and strategies for avoiding common failure points. Throughout, it offers concrete examples from The RAFT-supported work and concludes with practical advice for communities seeking to partner with colleges and universities, as well as guidance for academic partners aiming to structure their work in ways that are responsive, equitable, and useful.

Partners in the climate resilience ecosystem

Resilience planning benefits from partnerships and collaborations across sectors, scales, and knowledge systems. Boundary-spanning organizations play a critical role in this work by helping translate scientific information into decision-relevant knowledge, aligning funding with local priorities, and facilitating coordination among agencies, communities, and experts.

Federal agencies such as the National Oceanic and Atmospheric Administration (NOAA), Federal Emergency Management Agency (FEMA), and the U.S. Army Corps of Engineers provide data, guidance, and funding related to hazards, ecosystems, and infrastructure. State agencies often administer grants, manage coastal programs, and provide regulatory oversight. Nonprofit and community organizations may focus on conservation, community development, environmental justice, or climate adaptation, offering technical assistance and advocacy. Extension services, particularly those associated with land-grant universities, have a long history of working directly with communities to share applied knowledge.

These partners bring technical expertise, regulatory authority, funding, and facilitation capacity. Most coastal communities are accustomed to working with most of these partners. They know where to go for flood maps, grant opportunities, or regulatory guidance. They may have standing relationships with regional planning commissions or watershed organizations. These partnerships are essential, but they do not always meet all community needs. Agencies and nonprofits may be constrained by funding cycles, staffing limitations, or narrow programmatic mandates. They may not have the flexibility or capacity to engage deeply with individual communities over extended periods or to explore innovative or locally specific solutions. Communities seeking partners for resilience efforts should keep these limitations in mind and strategically create a partnership network that addresses their needs, realistically taking into account what each partner can offer and what they cannot. For example, The RAFT's three university partners organize and spearhead the effort, conduct the resilience scoring and the regional community workshops, and create the partnership network needed to implement a community's selected resilience actions. However, The RAFT university team itself cannot accomplish all of the resilience projects selected by a community, so we approach and coordinate with state agencies, nonprofit organizations, faith-based and other community organizations, and faculty at other universities to assist. Similarly, any community seeking to conduct a comprehensive resilience program should consider what they need, and then piece together the support system of partners that can best meet those needs in light of their available expertise, funding, and capacity.

Box 7.1 Boundary-spanning organizations

Boundary-spanning organizations act as the intermediary between the distinct cultures of mainstream science and local communities. These organizations are essential for fostering enduring collaborations, particularly when working with tribal nations and other populations that historically have been exploited by extractive research practices.[2]

Boundary-spanning organizations can also serve as neutral conveners, facilitating dialogue across silos, allowing representatives from government, industry, and the community to share perspectives and achieve a more complete understanding of complex issues.

Most coastal communities are already familiar with the following categories of boundary-spanning partners:

- Federal agencies (e.g., NOAA, FEMA, U.S. Army Corps of Engineers).
- State agencies, such as those administering coastal management, floodplain management, forestry, transportation, social services, health, and emergency management programs.
- Regional organizations (e.g., regional planning agencies and regional transportation agencies).
- Extension programs (e.g., cooperative extension, Sea Grant extension, family and consumer sciences, Master Gardener and Master Naturalist programs, and 4-H extension services).
- Nonprofit and community organizations such as watershed organizations, environmental nonprofits, and neighborhood associations.

University researchers, staff, and students can play boundary-spanning roles that build on their expertise. The RAFT itself demonstrates a successful model of a multi-university, multidisciplinary partnership that supports resilience planning and implementation by providing objective assessments of resilience, convening stakeholders from across the whole community, and facilitating implementation teams that advance resilience actions. Working with under-resourced communities across Virginia's diverse geography, we have been able to connect localities with university partners who provide critical support through faculty research and student projects. For example, through the application of The RAFT across different communities, university students, faculty, and staff have assisted coastal communities by conducting resilience assessments that are objective and customized to local needs. They have also been instrumental in identifying and implementing small-scale projects and providing research to support the development of plans that support resilience (see Chapter 4 for some examples).

Universities occupy a unique position in the resilience ecosystem. They sit at the intersection of research, education, and public service, and they often have mandates – formal or informal – and resources to engage with surrounding communities. Colleges and universities occupy a niche that draws upon:

- Long-term institutional presence.
- Respect based on serving in an objective, science-based role rather than engaging in advocacy, like nonprofits.
- Broad and deep disciplinary expertise.
- Students seeking applied learning experiences.
- Faculty and staff seeking to engage with communities in their research, teaching, and/or service.

Despite these assets, universities are rarely positioned as core components of the resilience support ecosystem. Understanding why requires examining both how universities are structured and how communities perceive them.

Universities as overlooked resources and partners

Colleges and universities occupy a complex and sometimes ambiguous role in society. On the one hand, they are centers of knowledge production, education, and public service. On the other hand, they are large institutions with their own priorities, bureaucracies, and power dynamics. For many community leaders, universities may seem distant, intimidating, or misaligned with local needs and goals. What's more, the community may even view the university as a powerful economic and political force that works against or undermines the community's needs and goals, leading to the proverbial town-and-gown tensions.

From the community perspective, universities can be difficult to navigate. Expertise is dispersed across departments, schools, and campuses, and there is rarely a single, obvious point of entry. A community seeking help with flooding, housing, or emergency preparedness may not know whether to contact a civil engineering department, a planning program, a public policy school, a law school, or a coastal science institute. And, even if they knew the appropriate entity, they may not know who at that program, department, or institute to contact and how.

There can also be trust issues. Many communities, particularly under-resourced or marginalized communities, have experienced extractive research relationships in which data was collected and research conducted, but little tangible benefit remained after a grant ended, students graduated, or faculty moved on. Worse, some populations have experienced life-changing and life-long harm from unauthorized use of their bodies for medical experimentation (such as harmful radiation or deliberate exposure to life-threatening viruses), or unauthorized surgeries (such as sterilizations). These long and deep histories in university communities can engender understandable skepticism, distrust, and active anger regarding the university's intentions. Another common historical source of town-gown tension arises from the university's ability to raise funding from its alumni for building new dormitories, classrooms, laboratories, and programs – which often encroaches on the surrounding community's land use hopes and plans, creating real estate inflationary pressures, pinch points, and resentment. Additionally, even if there were interest in connecting with the university, under-resourced communities may lack the time or staff capacity to manage university partnerships.

From the academic side, faculty, staff, and students also face structural barriers to partnering with communities on resilience efforts:

- Community-engaged work may not be recognized or valued by some programs, departments, or institutes or centers.
- Applied work often requires more relationship building, coordination, and time than traditional research. Faculty, staff, or students may not have the capacity or resources to build the necessary relationships and coordinate within the timeframe of a fall or spring semester.
- It also may be difficult to complete a community's requested project within one semester because students are not seasoned professionals in their field, and their learning curves must be accommodated. This requires that either the project is scoped for what one class can accomplish in a semester, often far less than what is desired, or the communities must be patient to obtain results, and that faculty must be willing to retrain a new group of students each semester.

Our experience as university faculty and staff, and through the application of The RAFT, demonstrates that these barriers are real. However, they are not insurmountable. If university-community partnerships are intentionally structured and supported, and time is allowed for the necessary relationship-building and coordination, then the partnership can be a worthwhile, productive, and effective experience for all involved. This can be achieved by coordinating in advance of the semester, or if a project is recognized by all at the outset as needing to span more than one semester. Moreover, university engagement practitioners can facilitate relationships between communities and faculty that last year after year, even if the student population that provides support changes each semester.

The RAFT partnership: A model that evolved with practice

The RAFT was established as a multi-university, multidisciplinary partnership. From the outset, The RAFT was designed as a collaborative platform linking universities, practitioners, and communities. One of the most significant contributions of The RAFT has been its role as an intermediary between communities and universities. Rather than expecting communities to navigate the complexity of academic institutions on their own, The RAFT became a translator and connector, matchmaking communities in need with faculty who have expertise and capacity to support them.

In practice, this has meant spending extensive time with community leaders to understand their priorities, constraints, and timelines. It also has meant working internally within universities to identify faculty and students whose interests aligned with those needs. In some cases, this involves adapting existing courses or research agendas. In others, it means creating new opportunities for engagement, such as internships or independent study projects outside of the normal academic calendar. For under-resourced communities, this kind of facilitation has been critical. Community staff often lacked the capacity to draft detailed project scopes, manage student teams, or supervise academic deliverables. The RAFT helped bridge these gaps by providing structure, coordination, and ongoing communication while investing in building trusted relationships on both sides of the partnership.

This intermediary role also helps manage expectations on both sides. Communities are able to articulate what they need and what they can realistically contribute, whether in terms of staff time, data, or local knowledge. University partners gain a clearer understanding of the practical constraints facing communities and the importance of producing usable, timely outputs, while providing students, often future resilience professionals, with high-impact experiences in how to work meaningfully in ways that support community-driven goals.

Engagement models for how universities can support resilience

There are several engagement models for how university faculty, staff, and students can contribute to resilience planning and implementation. Each has different strengths, limitations, and implications for community capacity (see Table 7.1 for four common engagement models). Regardless of the project's nature or model, it is vital that all members of the project team understand the community's historical context, strengths, and needs.

Common barriers to university-community partnerships

While the potential value of university-community partnerships is significant, our experience underscores that these collaborations are rarely straightforward. Some of the same barriers that challenge the establishment of university-community partnerships also impede successful collaboration.

Understanding these barriers in detail is essential for designing partnerships that are realistic, equitable, and effective. These barriers include:

1. **Capacity mismatch and uneven burden** – One of the most persistent barriers is a mismatch in capacity between university partners and community partners.
   From the community perspective, working with a university partner often requires:

   - Staff time for meetings, onboarding, and review.
   - Data compilation and sharing.
   - Responding to student questions.
   - Explaining local context, history, and politics.

   For communities with limited staff, this partnership work competes directly with day-to-day responsibilities. Even well-designed

Table 7.1 Four engagement models for university-community resilience partnerships

Class projects

Often the easiest entry point for university engagement.

Examples:

- Law, policy, and coastal resilience classes complete resilience scoring for localities using The RAFT Scorecard.
- Environmental policy and law classes conduct policy analyses and write white papers.
- Planning or environmental engineering classes develop maps and green infrastructure plans.
- Landscape architecture, architecture, or engineering classes develop preliminary designs for nature-based solutions.

Tip: Successful class projects have clearly defined scopes, align with course objectives, set realistic expectations, have a designated community contact, and provides sufficient time for students to learn about the community and its needs before engagement.

Caution: Class projects are constrained by semester timelines and student experience levels. Communities should view projects as exploratory/preliminary and not substitutes for professional services.

Internships

Offer more sustained capacity support and can be particularly valuable for under-resourced communities.

Can last a semester or longer. Generally has a minimum time commitment (usually 120–300 hours). Interns can be embedded directly within local organizations.

Examples:

- GIS (geographic information system) mapping of flooding and stormwater vulnerability
- Grant research and writing
- Creation of educational activities on climate, emergency preparedness, or coastal ecosystems.
- Wetlands monitoring and conducting vegetation surveys, erosion measurements, and water quality sampling.

Tip: Internships work best when students are adequately supervised and organizations have the capacity to integrate them meaningfully into ongoing work.

Caution: Expectations should be specific and clearly defined by the supervisor and communicated to the intern. Regular external oversight and check-ins are needed to ensure the relationship is progressing well for the intern and community organization.

Applied research projects

Involves faculty and/or students conducting research that addresses specific, practical questions posed by communities.

Table 7.1 (Continued)

Examples: • Flood scenario modeling for specific neighborhoods. • Evaluation of adaptation options. • Social vulnerability analysis. • Survey of residents about emergency preparedness or well-being. • Legal research on specific policy questions raised by the community. ***Tip:*** Applied research is most valuable when questions are co-developed with the community and findings are translated into actionable insights. ***Caution:*** Benefits can be lost or undermined if the research team does not build in time for training the community in how to use the research product or findings and/or adapting it for community use. For example, if a new flooding model is developed, the team should build sufficient time into the project to learn who would use it, and to train them in its use.
Participatory research projects Involves the community in conducting research that will address specific, practical questions posed by communities. Training community participants in the research methodology is vital. Students may be paired with community participants or may be involved in training the community participants. Examples: • Data collection, such as recording "King Tide" flooding. • Conducting and recording interviews with community elders or residents about a particular history or specific experiences. • Conducting surveys or inventories across the community. ***Tip:*** Co-design the research process with community participants from the outset. Build in time and resources for training, trust-building, and shared interpretation of results. ***Caution:*** Training, coordination, and ongoing oversight and consultation are needed. Funding for the community participants should be considered. Students must be provided with necessary training for project work.

projects can feel burdensome if they require sustained attention without immediate payoff.

From the university perspective, this burden is often invisible. Faculty and students may view community engagement as real-world learning, while community partners experience it as another obligation layered onto already stretched capacity.

2. **Mismatched timelines and the academic calendar** – Timeline misalignment can be one of the most common and predictable points of friction. Timelines rarely align. A community may need

information within three months to meet a grant deadline, while a class project may not conclude until the end of a semester or may pause entirely during academic breaks. Even short gaps in momentum (e.g., summer break) can impede progress, especially when staff turnover or emergencies occur.

3. **Student and staff turnover** – Much of the hands-on work in university-community partnerships is done by students. While this provides valuable learning opportunities, it also introduces a risk of knowledge attrition when students leave the university.
   When students graduate:

   - Institutional memory disappears.
   - Relationships must be rebuilt.
   - Communities may be left with unfinished work or undocumented decisions.

   Communities often report frustration when a new student team arrives with little understanding of prior work, forcing staff to re-explain context and revisit earlier decisions. This need to restart disproportionately affects under-resourced communities that have less capacity to repeatedly onboard new partners.

   - Similarly, local governments and tribes can also experience staff turnover, especially in under-resourced areas. This can be challenging if turnover occurs in the middle of a semester or class project. Or, if new staff are hired shortly after a class project is completed and a mechanism for sharing final products and data in an accessible way is not established, the class output may be lost to new staff. Thus, planning for institutional knowledge retention within both the local government and the university, beyond a single staff, faculty, or student, is critical.

4. **The incentive disconnect within academia** – A structural barrier lies in how academic success is measured and rewarded.
   Faculty are typically evaluated based on:

   - Peer-reviewed publications.
   - Grant funding.
   - Teaching evaluations and/or tenure reviews.

   Community-engaged resilience work, by contrast, often prioritizes:

   - Local relevance.
   - Timely outputs.
   - Longer-term relationship-building.
   - Incremental progress toward implementation.

These priorities do not always align. Faculty may struggle to justify the time required to maintain community partnerships when the work does not clearly advance traditional academic metrics. Students may be enthusiastic, but faculty support is essential for continuity and quality. It's important to recognize that even faculty who value engagement can be constrained by institutional expectations. We often hear early-career faculty say, "Wait until I get tenure. Then I will have more ability to take on these projects."

5. **The hand-off mentality** – A particularly challenging barrier is the assumption that a project is complete once a report, model, map, or plan is delivered. This hand-off mentality treats knowledge production as the endpoint rather than a step in a longer implementation process, reaffirming the view that universities simply "study and leave."

   University-community partnerships can be unproductive when the output handed off to communities is not actionable. These include:

   - Lengthy reports without specific prioritized recommendations.
   - Technical findings without guidance on next steps.
   - Recommendations that exceed local capacity or authority.

   Without additional support, these products are often not used, not because they lack value, but because communities lack the resources, knowledge, or connections to act on them.

6. **Administrative and legal friction** – Partnerships can be slowed or derailed by administrative hurdles, including:

   - Data-sharing agreements.

Table 7.2 Community and university timelines

Communities operate on timelines driven by:	*Universities operate on:*
• Grant deadlines • Fiscal years • Disaster recovery windows • Political cycles • Construction seasons • Economic seasons • Staff turnover and institutional memory gaps	• Fixed academic semesters or terms • Summer and winter breaks • Student graduation cycles • Grant deadlines • Student funding timelines • Multi-year research funding timelines

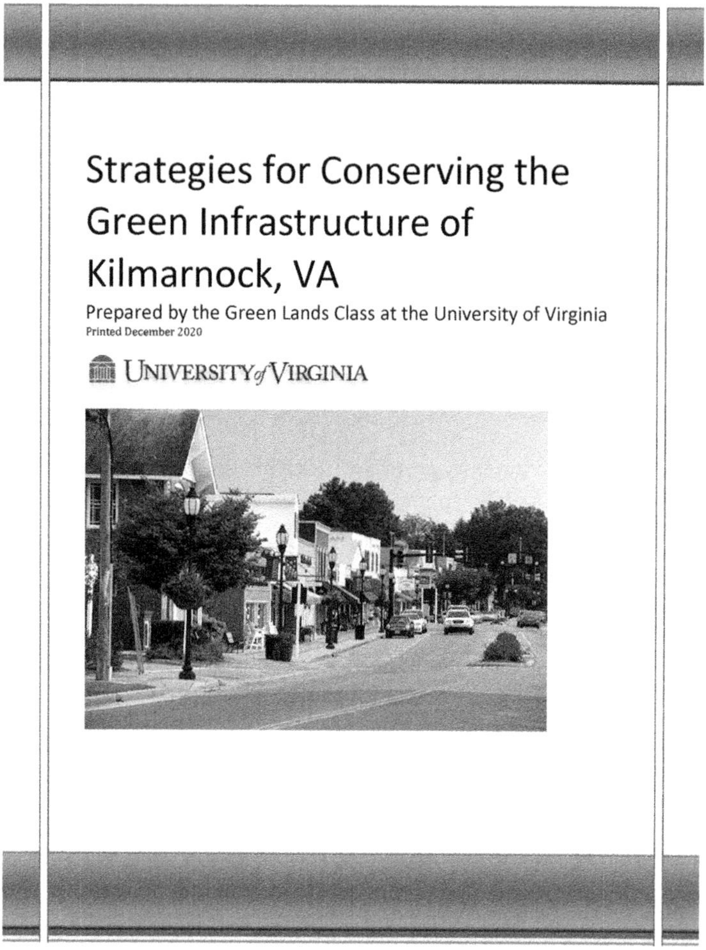
Strategies for Conserving the Green Infrastructure of Kilmarnock, VA

Prepared by the Green Lands Class at the University of Virginia

Printed December 2020

UNIVERSITY of VIRGINIA

Figure 7.1 Example of a class project where students developed strategies for conserving green infrastructure in a locality.

Source: Cover page of report by students in the University of Virginia's School of Architecture and supported by the Green Infrastructure Center Inc. Used with permission.

- Human subjects research and institutional review board (IRB) requirements.
- Procurement and contracting rules.
- Intellectual property concerns.

While these issues are often invisible at the outset, they can significantly affect timelines and feasibility.

Taken together, these barriers explain why many university-community partnerships fall short of their potential. They also reinforce this chapter's central argument: partners need help in serving communities. Without intentional design, support, and boundary-spanning facilitation, as well as the development of an understanding among university leadership of the need for this type of applied research and field work, even well-resourced universities and motivated communities can struggle to work effectively together. Recognizing these barriers is not about assigning blame. It is about creating realistic, respectful, and productive conditions for collaboration.

Making partnerships work

Colleges and universities are valuable but often underutilized partners in climate resilience planning and implementation. The RAFT experience demonstrates that, when partnerships are intentionally structured, grounded in community priorities, and supported by boundary-spanning intermediaries, universities can provide meaningful capacity and expertise, particularly for under-resourced communities.

At the same time, effective partnerships require effort, humility, and follow-through. When done well, they strengthen not only resilience outcomes but also the relationships and capacities that sustain resilience over the long term. To foster productive and sustained university-community collaborations, the university team needs to create with their community partners a strong foundation of deliberate planning, clear communication, and shared vision.

- **Build trust and shared vision**
  - Build trust before partnering – Successful partnerships are built on a foundation of trust and relationship building, which should begin before a specific project is needed.
  - Be accessible – University faculty and staff should be visible, actively serving on community committees and attending local

events, while inviting community leaders to the university. Securing a budget for in-person travel to meet with local government or tribal leaders can be critical.
    - Mutual understanding – Prioritize achieving a mutual understanding of needs and working toward a common goal.
    - Mutual respect – Acting with mutual respect is at the core of the partnership.
- **Establish clear expectations and scope**
    - Formalize expectations – A successful project must have stated expectations. This should result in a written document that lays out the responsibilities, roles, and objectives for both the university and community partners, ensuring clarity and mitigating confusion.
    - Objective scoping – Clearly define the project scope, focusing on bite-sized goals that are feasible within the project's timeframe.
    - Be upfront and honest – Be honest about capacity, needs and interests, and how/when things should happen to set realistic expectations.
- **Ensure effective communication and capacity support**
    - Communicate consistently – Establish agreed-upon communication channels and strategies to move forward effectively.
    - Address issues as soon as possible – Any issues or problems that arise during the planning process must be addressed immediately to maintain project momentum and preserve relationships.
    - Train students appropriately – Students need detailed training that covers professionalism, representation of the university, sensitivity to community needs and challenges, and the community's organizational context before they begin work. Community partners should be viewed as co-educators in this process.
    - Align timelines – University partners need assistance in structuring their work to meet community needs on community timelines. It is important to acknowledge that not every academic project is feasible within a single semester.

We offer the following guidance, based on our experiences working with communities, for making university–community partnerships work in advancing equitable resilience.

For community partners:

- Start with clearly defined needs.
- Match the engagement model to the problem.
- Scope projects realistically.
- Plan for staff time and coordination.
- Build relationships.
- Communicate regularly, and don't assume that university partners know details about your community.

For university partners:

- Listen first.
- Recognize, respect, and build on community strengths and capacity.
- Learn as much as you can about your partner community and its history before commencing work there, to help you understand community dynamics, past experiences, and historical traumas with legacies of harm.
- Translate findings clearly into impacts and actions.
- Plan for project continuity and implementation.
- Be honest about constraints, including timelines and required approvals for projects.
- Communicate regularly.
- Honor community privacy and teach students to exercise discretion and respect.

Box. 7.2 The bottom line for communities

University partnerships are most effective when they are community-driven, tightly scoped, realistically timed, and intentionally supported. Boundary-spanning organizations, clear expectations, and early implementation planning are often the difference between a helpful collaboration and a frustrating one.

Table 7.3 Community organization checklist: Deciding on whether, how, and when to partner with a college or university on a resilience project

Clarify the community need first

☐ Is the problem clearly defined in plain language (not academic terms)?
☐ Is this a capacity gap (e.g., staff time, analysis, documentation) or a technical gap (e.g., modeling, assessment)?
☐ Is there a near-term decision, deadline, or funding opportunity this work needs?
☐ Would a good enough product delivered on time be more valuable than a perfect product delivered later?

If you cannot answer these questions clearly, pause before engaging a university partner.

Match the engagement model to the need

☐ Class project or service learning → exploratory work, assessments, policy scans.
☐ Internship → filling specific, sustained capacity gaps.
☐ Applied research → decision support, modeling, evaluation of options.
☐ Mixed model → longer-term engagement with hand-offs planned in advance.
☐ Participatory research → data collection by trusted community members.

Avoid defaulting to the most intensive model. Start with the lightest approach that meets the need.

Scope the work realistically

☐ Can the project be completed within a semester or a clearly defined time window? Or is there capacity for a class in the next semester to pick up and build upon the work? Keep in mind that semester projects often are lined up by faculty many months in advance.
☐ Are deliverables limited, concrete, and clearly described?
☐ Is there a single point of contact on the community side?
☐ Have you explicitly said what the project will not do?

Overly ambitious scoping is the most common cause of frustration in university partnerships.

Address capacity and burden up front

☐ Do you have staff time to meet, review drafts, and answer questions?
☐ Have you identified what the community will contribute (time, data, local knowledge) and whether you have the capacity to provide these inputs?
☐ Have you minimized requests that pull staff away from core responsibilities?
☐ Is there an intermediary who can help manage coordination? Can you as a practitioner assist in providing some of the administrative support (scheduling meetings, facilitating discussions, taking notes, and following up on action items) to mitigate the lift?

A "free" project is not free if it overburdens limited staff capacity

Table 7.3 (Continued)

Align timelines early ☐ Does the academic calendar align with your planning or funding timeline? ☐ Are there known gaps (summer, spring, winter break, holidays) that need workarounds? ☐ Have you planned for continuity if students graduate or leave? Have you determined whether faculty members can commit to collaborating multiple semesters in a row or each year? **Assume timeline misalignment unless proven otherwise. Plan accordingly.**
Set expectations in writing ☐ Is there a short, written scope, Memorandum of Understanding (MOU), or service-learning agreement? ☐ Are roles, responsibilities, and deliverables clearly stated? ☐ Is there agreement on how results will be shared and used? ☐ Have match contributions (faculty or student time) been documented if relevant? **Written clarity protects both the community and the university partner.**
Prepare students and faculty for community work ☐ Do students understand the community context and constraints? Are they and the faculty willing to invest in traveling to the community, spending time with local leaders or staff, and listening to your interests, assets, and needs? ☐ Have expectations around professionalism and communication been set? ☐ Are community partners treated as co-educators, not just data sources? **Good preparation reduces rework and builds trust.**
Plan for the hand-off from the start ☐ Will the final product include a prioritized action list, not just analysis? ☐ Is there a clear answer to: What happens next? ☐ Have follow-up presentations or briefings been scheduled? ☐ Is knowledge being captured so it survives staff or student turnover? **A project is not finished just because the report, data, or map is delivered.**
Focus on use, not just output ☐ Can the deliverables be reused for grants, plans, or public meetings? ☐ Are results translated into plain language and visuals? ☐ Has the community confirmed the products are usable? **If it cannot be used, it does not build resilience.**
Reflect and decide what to do next ☐ Did this partnership reduce burden or increase it? ☐ Did it move the community closer to project implementation? ☐ Is this a relationship worth deepening, or should it stay project-based? **Not every partnership needs to become long-term. Intentional endings matter.**

Table 7.4 University checklist: Working effectively with communities on a resilience project

Start with community priorities, not academic ideas

☐ Has the community clearly articulated the problem or need?
☐ Are you responding to a real decision, deadline, or capacity gap?
☐ Have you resisted reframing the problem to fit your preferred topics and methods?
☐ Can you explain the project's value in plain, non-academic language?
☐ Have you explored how the project can recognize and build on community strengths and knowledge?

If the project primarily advances only your research agenda, pause and reassess.

Choose the right engagement model

☐ Is this best suited to a class project, internship, applied research, or mixed model?
☐ Does the model match the community's timeline and capacity?
☐ Are you transparent about what this model can and cannot deliver?

Do not oversell what a semester, a student team, or limited funding can accomplish.

Scope the work to be achievable and useful

☐ Are deliverables limited, concrete, and clearly defined?
☐ Is the scope realistic given student experience and time available?
☐ Have you explicitly defined what is out of scope?
☐ Would you be comfortable using these outputs yourself in practice?

A smaller, usable product is better than an ambitious one that never gets used.

Respect community capacity and burden

☐ Have you minimized demands on community staff time?
☐ Are meetings purposeful and scheduled with awareness of local constraints?
☐ Have you accounted for the emotional and political labor communities often carry?
☐ Are you compensating partners where appropriate or reducing unpaid labor?

Communities are not teaching labs; their time is not free.

Align with community timelines (not just the academic calendar)

☐ Do you understand the community's planning, funding, or political timelines?
☐ Have you planned for breaks, semester timeframes, and student turnover?
☐ Is there a continuity plan if students graduate or staff members change?

Assume misalignment unless you intentionally design around it.

Table 7.4 (Continued)

☐ **Put expectations in writing**
☐ Is there a written scope, MOU, or service-learning agreement?
☐ Are roles, responsibilities, and deliverables clearly documented?
☐ Have data ownership, use, maintenance, and attribution been discussed?
☐ Have match contributions (faculty/student time) been documented if relevant?
Written agreements protect relationships, not just institutions.

Prepare students for community-based work
☐ Have students been trained in professionalism and respectful engagement?
☐ Do they understand the community context, history, and constraints?
☐ Have you discussed power dynamics, discretion, and ethical responsibilities?
☐ Do students understand their role as ambassadors for the university in that their actions will shape and impact participant views of the university?
Students need preparation, not just enthusiasm.

Translate, don't just deliver
☐ Are findings presented in plain language with clear visuals?
☐ Have you avoided jargon and unnecessary technical details?
☐ Can a non-expert quickly understand the key takeaways?
If a community cannot explain your findings to others, the work is not finished.

Plan for implementation and the hand-off
☐ Does the project end with a prioritized action list?
☐ Have you clarified what happens after the final deliverable?
☐ Are you available for brief follow-up support (e.g., presentations, grant application help)?
☐ Is knowledge captured so it survives student or staff turnover?
☐ Set in place a mechanism for sharing all deliverables and outputs, ideally including raw data and a detailed description of methods, so that future staff will be able to access, understand, and use the products created.
Reports do not implement themselves.

Reflect on impact and reciprocity
☐ Did this partnership provide tangible benefit to the community?
☐ Did it reduce burden or inadvertently add to it?
☐ Were community partners credited appropriately?
☐ Should this relationship deepen, pause, or conclude intentionally?
Not every partnership must continue, but all should end responsibly and respectfully.

Lessons learned for practice

1) **Relationships matter more than products** – The strength of the relationship between community and university partners has consistently mattered for resilience projects more than the final product. Trust, clear communication, and mutual respect are proven foundations for useful work. Communities are more likely to use and act on outputs when they trust the people who produce them and feel heard throughout the process. Strong relationships also create flexibility. When timelines shift, data proves incomplete, or deliverables need to change, partners who have invested in communication are better able to adapt without damaging the collaboration. Conversely, technically strong products produced without trust are often unused or undervalued.
2) **Not all university engagement is the same** – One of the most common mistakes in university-community partnerships is assuming that all forms of engagement are interchangeable. The RAFT experience shows that different needs call for different engagement models and that choosing the wrong model can undermine even well-intentioned efforts. Choosing the lightest-weight model that meets the community's needs reduces burden, manages expectations, and increases the likelihood of follow-through. Overly complex engagement models can strain community capacity and slow progress, particularly in under-resourced settings.
3) **Intermediaries add real value** – A recurring theme across university-community partnered projects is the importance of intermediaries such as The RAFT or other invested practitioners who remain engaged beyond a semester or class timeline. The RAFT team has played a critical role in translating between academic and community worlds, aligning expectations, and reducing friction. Intermediaries help communities articulate needs in ways that university partners can respond to and help university partners structure projects that are feasible and useful. They also support coordination tasks that neither side is well-positioned to handle alone, such as matchmaking and scoping.
4) **Timeline alignment is essential and requires effort** – Timeline misalignment can be one of the most persistent and predictable barriers to effective collaboration. Community needs are often driven by grant deadlines, fiscal years, political cycles, and

hazard events. University timelines are usually shaped by academic semesters, student availability, and research funding cycles. These rhythms rarely align without deliberate planning. Our experience shows that timeline misalignment should be assumed, not discovered mid-project. Successful partnerships address this early by openly discussing constraints, planning around academic breaks, and building in buffers for delays.

5) **Equity requires intentional design** – Equity must be intentionally built into how collaborations are structured and resourced. Under-resourced communities often face the greatest risks while having the least capacity to engage in complex partnerships. Without care, university collaborations can unintentionally place additional burdens on these communities. Work that appears free on paper often carries hidden costs in staff time, emotional labor, and administrative effort. Successful partnerships explicitly acknowledge these costs and work to minimize them by narrowing scopes, reducing meeting demands, and providing coordination support.
6) **Implementation must be planned from the start** – Perhaps the most critical lesson from The RAFT is that reports alone do not build resilience. Many communities already have plans, assessments, and studies. What they often lack is the capacity to move from analysis to action. Effective university partnerships are planned for implementation from the beginning. This includes providing prioritized recommendations, identifying feasible next steps, and supporting transitions beyond the life of a class. In some cases, this has meant presenting findings to elected officials or assisting with follow-on funding applications. For example, the creation of a small business resilience assessment tool for The RAFT was followed by workshops to help small businesses use the assessment and develop strategies to improve their resilience.

Notes

1 Denckla Cobb, T., & Bingham, S. (2024). The art of the town-gown dance: Healing legacies of harm in our food system through equitable pedagogy. In S. Raja, M. Caton Campbell, A. Judelsohn, B. Born, & A. Morales (Eds.), *Planning for Equitable Urban Agriculture in the United States: Future Directions for a New Ethic in City Building* (pp. 453–487). Springer International Publishing.

2 Schwartz, M. W., Fleishman, E., Williamson, M. A., Williams, J. N., & Morelli, T. L. (2021). The use of boundary-spanning organizations to bridge the knowledge-action gap in North America. In C. C. Ferreira & C. F. C. Klütsch (Eds.), *Closing the Knowledge-Implementation Gap in Conservation Science: Interdisciplinary Evidence Transfer Across Sectors and Spatiotemporal Scales* (pp. 229–254). Springer International Publishing.

Index

For Product Safety Concerns and Information please contact our EU representative GPSR@taylorandfrancis.com Taylor & Francis Verlag GmbH, Kaufingerstraße 24, 80331 München, Germany

Batch number: 10569667

Printed by Printforce, the Netherlands